Wildlife Management in Savannah Woodland

Originally published in 1979, *Wildlife Management in Savannah Woodland* provides a multidisciplinary approach to the environment. Developed by local scientists with a deep knowledge and understanding of the local situation, the book provides a pragmatic and realistic approach to West African conditions.

Wildlife Management in Savannah Woodland

S.S. Ajayi and L.B. Halstead

Routledge
Taylor & Francis Group

First published in 1979
by Taylor and Francis Ltd

This edition first published in 2018 by Routledge
2 Park Square, Milton Park, Abingdon, Oxon, OX14 4RN
and by Routledge
711 Third Avenue, New York, NY 10017

Routledge is an imprint of the Taylor & Francis Group, an informa business

© 1979 Taylor and Francis Ltd

Publisher's Note
The publisher has gone to great lengths to ensure the quality of this reprint but points
out that some imperfections in the original copies may be apparent.

Disclaimer
The publisher has made every effort to trace copyright holders and welcomes
correspondence from those they have been unable to contact.

ISBN 13: 978-0-8153-6628-7 (hbk)
ISBN 13: 978-1-351-25924-8 (ebk)
ISBN 13: 978-0-8153-6636-2 (pbk)

Wildlife Management in Savannah Woodland

Recent progress in African studies

Edited by
S. S. Ajayi and L. B. Halstead

Co-Editors
C. Geering, J. B. Hall and L. P. van Lavieren

TAYLOR & FRANCIS LTD
10-14 Macklin Street, London WC2B 5NF
1979

First published 1979 by Taylor & Francis Ltd,
10-14 Macklin Street, London WC2B 5NF

©1979 Taylor & Francis Ltd

ISBN 0 85066 175 7

Production services by Book Production Consultants,
7, Brooklands Avenue, Cambridge

Contents

Section 3. Fire

Section 4. Management, Training and Education

PREFACE

Wildlife management in West Africa is concerned primarily to ensure the rational exploitation of the natural resources of the region. The approaches and perspectives contrast markedly with those of the past decades which have characterized the East African situation. Tourism is a major industry in East Africa attracting as it does large numbers of visitors from North America and Europe. Its success, however, is dependent on the vagaries of the economic climate of distant lands. A reduction in the level of affluence of advanced industrial nations can have catastrophic effects on a country that relies unduly on foreign tourists for a large proportion of its income, with serious consequences for the survival of its wildlife.

The long-term strategy being evolved in West Africa holds out a greater hope for the survival of wildlife throughout the continent. The competing claims on the resources of the local population and a tourist industry which characterize East Africa are not a source of conflict in West Africa. Indeed the entire rationale of wildlife management in West Africa is directed towards utilising wildlife as a renewable resource for the direct benefit of the local population rather than for the entertainment of wealthy foreign visitors.

To this end current research is being directed towards clearly defined objectives which have a practical application. The work being pioneered in West Africa, exemplified by the contributions to this volume, is characterized by the integrated multidisciplinary approach to the totality of the environment. The first prerequisite must be the collection of the basic ecological data, aspects of which are given in Section 1 *Population dynamics and monitoring*. Studies of the relationship between the fauna and flora are documented in Section 2 *Habitat utilization*. One area which generally receives little attention is the role of fire which is critical in savannah woodland and this is dealt with in Section 3 *Fire*. The final Section 4 *Management – training and education* is concerned with the practical application of knowledge to wildlife management: cropping game on a sustained yield basis as well as the recognition of suitable candidates for domestication.

This present volume heralds a new synthetic approach to wildlife an approach that has been developed by local scientists with a deep knowledge and understanding of the local situation a pragmatic and realistic approach that owes little to theoretical considerations proposed by others with no real experience of the West African conditions. It is anticipated that the contributions given in this volume will encourage comparable integrated approaches to the exploitation of forest resources in many other tropical regions in the world.

L. B. Halstead

INTRODUCTION

The Ibadan/Garoua International Symposium on Wildlife Management was a joint venture between the Department of Forest Resources Management, University of Ibadan, Nigeria and the College of African Wildlife Management, Garoua, Cameroun Republic. The international co-operation between the two institutions which are principally concerned with manpower training for the management of wildlife in Africa has matured not only into organizing this bilingual symposium (French and English) but also in combined student activities, planning and executing research projects which are of direct relevance to the development of wildlife in West and Central Africa.

The importance of this symposium is easily apparent. First, most of the national parks and game reserves in this region are in their early stages of development, especially when one considers development in terms of research and its application to the management of the ecosystem. Furthermore, a greater number of these reserves are situated in lowland savannah woodland areas. Most of the research methods developed in East and South African open rangeland, which has different ecological conditions, are of little application to the woodland areas of West and Central Africa. Areas in which research methods have to be developed for the savannah woodland to enhance wildlife management include censusing, population dynamics, and habitat analysis. The theme of the conference 'Wildlife Management in Savannah Woodland' was therefore very appropriate in view of these pressing problems. It was proposed that an international symposium of this kind could bring wildlife ecologists together, especially from Africa, to deliberate on and assess these problems. It would also provide an avenue to exchange ideas on experience gained by field workers and where relevant information on the theme of the conference can be gathered together, scrutinized and published as a first-hand information document or as working material for ecologists in West and Central Africa. It must be pointed out that West and Central Africa is a region of Africa which, though blessed with abundance and variety of wildlife, has been hitherto neglected compared with East and South Africa in terms of research and management.

About 40 papers were presented within the theme of the symposium. These were grouped into ten sessions, each tackling a specific topic with contributions from experienced field workers. It is recognized that wildlife cannot be managed in isolation from their habitats and vice versa. Therefore, throughout the symposium, emphasis was placed on the broad topics of population, habitats, the interaction between wildlife and their habitats, management planning, and manpower training. The ten sessions of the symposium thus represented specific topics within these broad areas. Population dynamics, censusing and population monitoring were three sessions which dealt principally with techniques used in age determination, indirect censusing techniques, large-scale total counts for monitoring large animal populations and predator–prey interaction. The section on habitats included sessions on the use of fire in habitat management. Fire is crucial in the evolution of the savannah, and remains the most important tool for the management of wildlife habitats. The session on fire attracted contributors who have actually worked in the savannah woodland ecosystems and valuable contributions were made on the structure, and productivity of savannah grassland in relation to grazing and burning. The sessions on the use of habitats by wildlife contained information on the ecological factors which influence the distribution of animals in nature and how this knowledge could be used for the management of wildlife populations. Contributions were also made on the habitat requirements and food preferences of certain wildlife species.

Manpower training was regarded as a major part of the symposium, first because it is necessary to draw attention to the acute shortage of wildlife management personnel in this part of Africa. Secondly, it was pertinent to point out the problems of wildlife management that are unique to West and Central Africa, and at the same time to try as much as possible to identify the relevant type of training to give to those who will manage this resource. It was also relevant at this point to learn from the successes as well as the mistakes of the experience gained by our East African counterparts in their manpower training.

The symposium succeeded in identifying problems of wildlife management that are unique to West and Central Africa and valuable information which was hitherto unknown to exist was brought together and this would certainly be of use to wildlife managers in this and similar regions of the world. It was very gratifying that the conference attracted a large number of participants from all over the world, particularly Africa, and because it was bilingual it represented a collection, perhaps of the largest wildlife technical experts in any African wildlife conference. It provided a valuable opportunity for wildlife experts from different localities to meet and discuss and exchange ideas on wildlife matters. This opportunity probably never existed before. The symposium also focused the attention of African governments and international organizations which deal with wildlife conservation on the potential of wildlife in West and Central Africa, and has highlighted the problems and gaps in the knowledge that exist in wildlife management. It was thus possible as a result of the symposium to identify our future priorities in research and management of the wildlife resources.

The present volume contains 26 selected papers that were given at the symposium.

The steering committee would like to record sincere appreciation to the Federal Government of Nigeria through the Agricultural Research Council of Nigeria and the Federal Department of Forestry without whose generous financial support, this conference could not have been held. Our profound gratitude is also due to the United Nations Food and Agricultural Organization (FAO) and the government of the Cameroun Republic that supported the organization of the symposium financially. The FAO also provided funds to enable experts in all their wildlife projects in Africa attend the conference. We are indeed grateful to the Rockefeller Brothers Fund and the United Nations Educational, Scientific and Cultural Organization for their financial contributions.

Our grateful acknowledgements go to the Staff (at all levels) of the Department of Forest Resources Management, University of Ibadan, Nigeria, and the College of African Wildlife, Garoua, Cameroun, for their selfless services in the Organization of the Symposium.

Finally it is our pleasure to pay tribute to Professor L. Roche whose recognition of the paramount need for an integrated approach led him to restructure the former Department of Forestry so that it now encompasses all aspects of Forest Resources Management, including as an integral part Wildlife Management. The convening of the symposium at Ibadan is a measure of Professor Roche's encouragement and support for the new integrated approach to forest resources that is being pioneered in West Africa.

S. S. Ajayi
C. Geerling
J. B. Hall
L. B. Halstead
L. P. van Lavieren

SECTION 1
POPULATION DYNAMICS AND MONITORING

1

WILDLIFE INVENTORY BY REMOTE SENSING TECHNIQUES

P. R. O. Kio

Department of Forest Resources Management,
University of Ibadan, Nigeria

INTRODUCTION

An increase in the size of human populations is almost always at the expense of wildlife populations, particularly of large animals. Widespread destruction of wildlife and its habitats precedes the establishment of sedentary agriculture and animal husbandry. In some developing countries of Africa, bush meat is the main source of animal protein amongst rural communities and the demand for it is rapidly increasing at the very time when the supply is diminishing. The imminent shortage of most kinds of game calls for the wisest possible management. This requires an accurate knowledge about the density and distribution of the surviving animals.

Management of any single resource frequently needs comprehensive information on other components of the environment and their interrelationship, in addition to the vital facts about its amount, quality and extent. To provide a firm ecological base for management, an inventory of a wildlife complex should consider the vegetation cover as an integral part of the environment and take into account its interdependence with landscape features, moisture conditions, physical characteristics of the land, distribution of water bodies, and past and present land-use practices (Gimbarzevsky, 1973). Inventories not only provide information on numbers and distribution and for determining the need for refuges, they also help to determine the length of hunting seasons, trophy limits, the need for closed areas and the special regulations required to restore depleted populations. Moreover, they give information for the management of good wildlife areas and on the extent and character of possible reduction measures that have to be made in areas over-populated with animals.

It is the dynamic nature of wildlife populations which has made their management such a difficult task. Management must be responsive to changes induced by environmental factors, human influences and natural events. Inventories need to be repeated at suitable intervals — a monitoring process in which remote-sensing techniques could well begin to play an important role. Some of these techniques have been in use in the last two decades in forest inventories and others are still very much at an experimental stage and at present find little practical use outside the laboratory. The aim of this paper is, therefore, to highlight some of these techniques in so far as they can be applied to the study and management of wildlife populations and to indicate trends in the development of the science, particularly with regard to an integrated and interdisciplinary approach to the solution of associated problems.

PROBLEMS OF WILDLIFE INVENTORY

As in forest inventory, the main objective for a wildlife inventory is to present estimates of the numbers of animals in an area according to a series of classifications such as species, sex, age, condition and habitat. The main problem is how to obtain reliable estimates of the varying populations with sufficient accuracy in order to meet management requirements. Complete census on a large scale, where it is at all possible, is usually out of the question on account of the cost. If estimates are to be based on a sample then accurate delineation of strata — relative densities, vegetation types, or niches of the different species — is crucial.

Obviously, there is no entirely satisfactory inventory method which can find universal application. Methods that seem to give good results in some areas may be inadequate in others where cover and terrain are different. Drive-counts, strip-enumerations, feeding-ground counts, forage-utilization and aerial-survey methods all are of value but no individual technique can be expected to produce completely adequate information. As an example, in the open savannah and treeless flood-plains aerial survey of large mammals is possible. The main difficulties lie with nocturnal animals and with large mammals whose feeding and watering activities are mainly confined to periods of dawn and dusk. In denser woodlands aerial surveys give far less accurate results.

The implication is that if large areas are involved, only a multi-stage system of inventory, which combines remote sensing with ground techniques, could yield quick and reliable estimates of wildlife populations. Under this system progressively more detailed information is obtained for progressively smaller sub-samples of the areas being studied. It may be possible in the not too distant future to identify and map most of the vegetation zones over a vast area on satellite photographs and to correlate these to animal concentration by using known relationships between vegetation types and animal densities. Given this limited information, a few representative areas should be studied more closely with a second stage of photography — for instance, large-scale, high-resolution aerial photography — to determine the relative or absolute abundance of the animals. Having ascertained the relative density of the animals, their watering-places and feeding areas, a third stage sub-sample may be visited on the ground to collect still more complete data on species distribution, sex and age ratios etc., for drawing up a comprehensive management plan for the whole area.

REMOTE-SENSING TECHNIQUES

Remote sensing is the study of objects from great distances. It involves the acquisition of data of distant phenomena or objects and their surroundings. The term is also used specifically to describe the techniques of sensing as a means of gathering information on vegetation, rocks, soils, water, wildlife and other components of the earth's environment (Kio, 1974). Every object emits or reflects rays from different zones of the electromagnetic spectrum. Sensors are installed in satellites or aeroplanes to collect data which give a measure of the reflectance or emittance within portions of the ultra-violet, visible, infra-red and microwave regions of the spectrum. The ray that reaches a sensor reveals the special properties of the object and remote-sensing techniques allow the reception, recording and evaluation of this information. A wide range of natural phenomena can be measured by these techniques and each part of the electromagnetic spectrum provides a specific type of information about the object being studied.

There are two main types of remote-sensing systems — photographic and non-photographic. The most extensively used are photographic sensors, which operate within the visible and the reflective, near infra-red region of the electromagnetic spectrum, that is, function within the range of 0.38 to 1.0 μm. Ordinary cameras can only perform within this range. From the lower end (less than 0.3 μm) visible light ceases. At values greater than 1.0 μm, lenses may resolve the 'light' but photographic films are no longer sufficiently sensitive and no image will be recorded. There are, however, experimental films which can operate between 1.1 and 1.2 μm. The photographic sensors consist of cameras equipped with high quality lenses, filters, and films to produce multi-spectral black and white, infra-red, colour or 'false' colour images. The non-photographic sensors consist of optical-mechanical scanners as well as non-scanning detectors ranging from video-cameras, multi-spectral and thermal scanners, to radiometer, radar, radio-sounding, magnetic- and gravity-sensing systems (Kio, 1974; Gimbarzevsky, 1973; Hempennius, 1969).

Most of these image-formation techniques are particularly suitable for sensing from a moving platform ranging from satellite to submarine, from helicopter to hovercraft, from boats and balloons to small-engine aircraft. Before a brief description is given of individual sensors likely to be of immediate or future use in the inventory of wildlife, it is essential to stress that remote sensing is a science in which hardware technology has far outstripped the ability of users to fully exploit the potentiality of old systems before assimilating the impact of new inventions. Several sensors which perform similar functions are undergoing simultaneous development.

Systems most suitable for resource mapping and measurement are conventional aerial photography, radar pictures, thermal and multi-spectral imageries obtained from Earth Resources Technology Satellites (ERTS — now called LANDSAT).

Conventional aerial photography

The conventional aerial survey cameras are used to provide aerial photographs for photogrammetric mapping and also for forest photo-interpretation. The usefulness of the photographs for wildlife inventory depends on the photo-scale and ground conditions. Small and medium-scale photographs are useful for describing and mapping the habitats. In open woodlands and grasslands large-scale photographs can be used to count big game animals.

The most commonly used film-filter combination in forest photograph interpretation is panchromatic film with a medium-to-deep-yellow filter. Colour photographs are increasingly used for specialized purposes. Infra-red colour films and the so-called false colours can be employed for identifying herds of animals and individual large mammals if taken on a sufficiently large scale. Infra-red film is particularly useful for separating dry and wet areas and therefore can be used for mapping the haunts of certain species of animals.

The main limitation of conventional aerial survey is that for accurate mapping, ground control points have to be established. The British Columbia Forest Service has developed a technique of low-level aerial photography of fixed base, which allows accurate photogrammetric measurements without ground control (Waelti, 1973). The equipment used consists of two Hasselblad Mark 70 cameras, mounted 4.5 m apart on a boom which is flown, suspended beneath a helicopter, 91 to 183 m above ground. Approximate flying height is controlled by a Bonzer radar altimeter to keep it reasonably constant in areas of uneven topography. The cameras are aligned to within 30 seconds of arc and their shutters synchronized to within a few milliseconds.

The resulting large-scale photo-pairs are used to measure resource elements that are not readily accessible but are important for compiling inventories and deciding resource-management techniques. The photographs are used for all kinds of forest inventory work, forest productivity studies where tree height and crown shape are measured periodically to assess the effect of silvicultural treatments and to develop growth simulation models. At an average height of 152 m above the ground large mammals are readily visible and a census can be taken without any difficulty. Smaller animals can be distinguished if the photographs are studied in stereomodel under a mirror-stereoscope equipped with high-power lenses. The method has been used for the study of fish habitats in streams, inventory of floating timber and debris on large reservoirs, and regeneration surveys.

Radar imagery

Unlike other remote-sensing systems which are passive, that is, which record the energy derived from the objects being sensed, Side Looking Airborne Radar (SLAR) is an active system since the energy is provided by the sensing device. This system was originally developed for military purposes but it is now finding increasing use in forest resource mapping. A narrow beam of pulsed microwave energy (radio signals) is sent to the earth from a transmitter mounted on board the aircraft. The intensity of the radiation reflected back by the object on the ground is picked up by a receiving antenna mounted on the aircraft. The quality and the intensity of the reflected radar signal depend on the surface characteristics and the physical and chemical properties of the target objects. Radar signals penetrate clouds, fog and rain and also, to some extent, vegetation (Kronberg, 1975), and are particularly useful in the tropics where large areas are covered by cloud for the greater part of the year and, therefore, are not accessible to conventional aerial photography.

Imageries from Earth Resources Technology Satellites (ERTS)

Until quite recently most scientific studies of the earth's resources have been accomplished from ground observations supplemented with aerial photographs taken from aircraft. Photographs obtained from satellites have made the study of resources on a world-wide basis possible. Space surveys could indicate the capability of different types of land and even keep track of the movement of locusts and other pests. Different grazing practices have been seen quite clearly from space (Colwell, 1973). Satellite photographs have provided the means of evaluating national grazing lands, and in the United States it is estimated (Pardoe, 1969) that the increase in cattle production of one per cent, which would be made possible by the additional grazing information, would pay for the whole of the research and development and launching of the ERTS programme.

A vertical photograph taken from satellite orbits 200 km above the earth can cover up to 20 000 sq. km. Conventional aerial photography of the same area would run into several hundred exposures. In no other way can a wildlife manager see up to 34 000 ha of land on one picture with uniform illumination. Despite the large area covered, satellite photographs are reported to have very good spatial resolution so that objects on the ground down to several metres in size can be detected, outlined and identified (Kronberg, 1975). For military and security reasons, such high resolution pictures are restricted and are not yet available for routine commercial distribution. When such photographs are eventually released for public use, satellite photographs will be directly competitive with, and probably more useful than, conventional aerial photographs, since a narrow satellite photograph is largely orthographic and can be used for preparing maps at a much smaller cost than that involved when conventional aerial photographs are used.

Due to the experimental nature of ERTS satellite, it has induced many workers from different disciplines and land-managing agencies to investigate ERTS data to solve specific problems. Heller (1973) has reported some surprising results from such studies. For instance, ERTS imagery over northern Alaska revealed major faults previously not discovered on aerial photographs, and research workers at Purdue University were able to discriminate crop types, water resources and soil boundaries in western Texas.

Imageries from satellites are mainly recorded by means of a scanning device called optical mechanical scanner. The sensor produces a nearly continuous image from a series of line scans. The radiation collected by the optical mechanical scanner is converted to electrical signals which can be directly beamed to satellite tracking stations on earth in the form of television pictures or which may be recorded on magnetic tapes within the satellite or the tracking station. The taped video data may be converted from magnetic tape to one or more of several types of signature data or imagery displays. When several spectral images covering the same territory are superimposed accurately in an image reproduction unit the 'false' colour picture which results from the original separate black and white transmission is extremely valuable for distinguishing various features of the terrain.

FUTURE DEVELOPMENTS

Multi-spectral scanning procedures from satellites could supply quantitative data in electronic form which can be interpreted automatically by computers. As an example, if the spectral 'signatures' of certain types of soil, rock, plants or animals are known, it is possible to program the computer to print out a map of all those areas of terrain in which radiation of a specific wavelength and intensity was recorded and therefore in which particular kinds of minerals, soil, plants or animals occur. In fact, appropriate computer programs are already available for digital mapping of crop distribution.

Computerization of photo-interpretation procedures is a major development which will revolutionize remote-sensing technology and make possible the preparation of maps tailored to meet specific requirements and for particular end uses. The main problem is the identification of the spectral characteristics of the materials to be mapped, though extensive studies are in progress in the United States and other countries to determine the reflective and emissive characteristics of natural materials.

Remote sensing is not only restricted by the sensing device but also by the physical properties of the atmosphere. Those parts of the atmosphere in which electromagnetic energy passes easily are usually known as 'atmospheric windows'. There is a very good window for aerial photography, two windows for thermal sensing, and a little investigated window at 20 μm and further windows for the microwave band used for radar.

The thermal sensing windows may be very useful for studying living objects on the earth's surface. Thermal radiation of an object varies with its temperature. Due to their specific physical and chemical properties objects on earth heat up differently during the day and cool off at different rates in the night. Thus temperature differences can be used to distinguish between various materials once the most favourable time for thermal mapping of particular objects is ascertained.

However, most studies in thermal sensing are at present confined to the temperate regions. No doubt once the system is perfected the technique could be adapted for use in warmer parts of the world. Since infra-red radiation can penetrate haze and clouds, infra-red thermography will be more useful than SLAR in that a small area can be imaged on a large scale and visual interpretation of the data is feasible.

REFERENCES

COLWELL, R. N. (1973) Remote sensing as an aid to the management of earth resources. *American Scientist*, Vol. 61, No. 2, pp. 175–83.

GIMBARZEVSKY, P. (1973) Remote sensing in the integrated surveys of biophysical resources. *(Proceedings Symposium IUFRO Subject Group S6.05., Freiburg).* Pp. 127–42.

HELLER, R. C. (1973) Analysis of ERTS imagery problems and promises for foresters. *(Proceedings Symposium IUFRO Subject Group S6.05., Freiburg).* Pp. 373–93.

HEMPENNIUS, S. A. (1969) Wall chart of image formation techniques for remote sensing from a moving platform. *(I.T.C. Publications, Series A46/B53*, Delft, Netherlands).

KIO, P. R. O. (1974) Developing countries and the new science of remote sensing. *Commonwealth Forestry Review*, Vol. 53, No. 2, pp. 137–45.

KRONBERG, P. (1975) Earth resources technology satellites (ERTS) and their application. *Applied Sciences and Development*, Vol. 5, pp. 68–77.

PARDOE, G. K. C. (June, 1969) Earth resources satellites. *Science Journal*, pp. 58–67.

WAELTI, H. (1973) Low-level, fixed base aerial photography for resources management. *(Proceedings Symposium, IUFRO S6.05. Freiburg).* Pp. 163–78.

2

THE CONCEPT AND PRACTICE OF ECOLOGICAL MONITORING OVER LARGE AREAS OF LAND: THE SYSTEMATIC RECONNAISSANCE FLIGHT (SRF)

M. D. Gwynne and Harvey Croze

UNDP/FAO, Kenya Habitat Utilization Project, Nairobi, Kenya

INTRODUCTION

Resource management organizations in East Africa, such as the range management divisions, game departments and national parks, require large-scale ecological data on which to base utilization and development programmes for extensive tracts of non-urban land. This demand has encouraged the gradual development within East Africa of the ecological monitoring concept — from the beginning when methods were sought to answer simple questions (e.g. How many animals are there?) to the present when complex land-use management questions are being posed.

The need for information on a large scale has led to the development of techniques for data gathering which are efficient and inexpensive enough to be applied repeatedly over large areas. Considering the difficulties caused by the dynamic state of rangeland ecosystems which experience both long- and short-term climatic cycles, the new methods of data gathering and data handling have been most successful (Norton-Griffiths, 1972; Cobb, 1975).

In Kenya, we have an approach whereby we look at the land's biological resources from three levels — from the ground, from the air and from space through the eyes of satellites. In this paper we will consider only the middle tier — aerial reconnaissance (Gwynne and Croze, 1975).

When considering monitoring strategy for a region about which little is known, the initial emphasis should be on the use of aerial techniques because aerial survey is the logical first field-step in the resource assessment of any new large development. The techniques used can provide useful quantified quick-look data at low cost (Watson, 1969). In any case, the deployment of more traditional ground techniques may be usefully considered as a function of the aerial strategy. Similarly, the interpretation of satellite imagery can be done in the spatial framework of the aerial reconnaissance.

The basis of the method is the ecosystem approach, the underlying aim of which is to determine the spatial and temporal pattern of primary and secondary productivity within a particular ecosystem or self-contained land unit.

The system of data collection discussed below can be applied at different levels of complexity from the simple to the comprehensive, and is as applicable to non-park areas as it is to the parks and reserves. What determines the level of application is the primary objectives of the survey, the funds and time available, the physical nature of the terrain, and the personnel that can be used in the survey and their technical skills.

A clear understanding of the requirements is absolutely essential to the planning of any aerial monitoring strategy: they determine the kind of ecological data that should be collected. In terms of costs and manpower it is as inefficient to collect too much data as it is to collect too few.

Care must be taken before a survey to ensure that the methods of data analysis proposed are compatible with the actual means and skills available. Un-analysed data are wasted data. It is no use becoming involved in a complex monitoring strategy dependent upon technical skills and computer facilities for data analysis that are simply not available to the project. In such a case it is more practical to attempt a much lower level of data collection and interpretation which will provide immediately useful ecological management information.

It is the purpose of this paper to suggest how simple aerial monitoring programmes may be planned for use in areas where long-term background habitat data are not available, and where trained scientists, technicians and analytical facilities are all in short supply.

AERIAL MONITORING

Information categories

Ecological monitoring uses information from three general categories:

(1) *Environmental:*   including information on climate, hydrology, topography soils and floristic dynamics.

(2) *Faunal:*   including information on wildlife and livestock numbers, distribution, population dynamics and habitat utilization.

(3) *Economic/political:*   including land-use forms, projected land demands, and national development goals.

The last category is normally supplied by the agency that has requested monitoring data. It will usually state what the projected land-use demands for the study area might be in relation to the national development goals. Depending on the political situation, the state of the economy, and the level of existing knowledge of the area, this statement of development intent can take several different forms. For example:

(*a*) It is government's intention to increase the livestock capability of the region.

(*b*) Recognizing the livestock production potential of the region, it is, nevertheless, government's intention to exploit fully the wildlife resource of the same region.

(*c*) The region is recognized as being one of great wildlife interest; it is government's intention to investigate the possibility of establishing within it one or more national parks as reserves to enhance the nation's tourist attraction potential.

(*d*) Government is uncertain of the development potential of the region and seeks advice on possible biological resource development strategies.

The form of the statement of development intent will make it clear to the investigators what the administrators see as the development ideal for the region. This, in turn, will help the survey team choose the appropriate habitat parameters for monitoring, paying particular attention to those which are necessary to determine whether or not the administrators' development ideas are viable.

Ecosystem boundaries

As the monitoring approach is essentially an ecosystem one, an initial step is to decide whether or not the area to be examined is in fact an ecosystem. Gazetted boundaries are almost invariably political rather than ecological so that they do not normally encompass organically and energetically self-sufficient units. It is necessary, therefore, to know whether the study area is part of one or more larger ecosystems, or contains one or more smaller self-contained ecosystem units, or, most rarely, is an ecosystem unit on its own.

This can be ascertained in general terms by taking the limits of movement of the largest animal biomass components as marking the approximate ecosystem boundaries (Pennycuick, 1975). Such animal movements are caused by various factors and it is part of the problem to determine these factors. To do this the ecologist has to describe the static structure of the ecosystem (topography, drainage, soils, vegetation, plant and animal community components), as well as to analyse the dynamics of the system (changes in time and space of climate and productivity, shifts in community structure etc.). If attempted in its entirety this can be a formidable task which can only be accomplished with careful planning and a systematic approach.

Measurable habitat attributes

The choice of a collecting or sampling strategy will depend on the spatial and temporal distribution of the phenomena being measured. It is convenient, therefore, to classify ecosystem attributes along a continuum of change from those which remain more or less constant over the long term to those which alter appreciably between sample periods, viz.

(i) *Permanent attributes:* topography, soils, drainage, water-holes, and static animal features such as termite mounds.

(ii) *Semi-permanent attributes:* plant physiognomy (cover vegetation type etc.), plant community composition zoogenic features (wallows, salt licks etc.), distribution of non-migratory large mammal species, and human settlement (villages, roads, farms, ranches etc.).

(iii) *Ephemeral or seasonal attributes:* rainfall, insolation, soil moisture, evapotranspiration, plant phenology, plant productivity (biomass, part composition, energy content etc.), distribution of migratory large mammal species, large mammal population, structure, fire and surface water.

Note that useful data on many of these attributes may be collected from our three operationally separate levels — from the ground, from the air, and from space via one of the earth resources assessment satellites.

Survey frequency and costs

The frequency of aerial surveys of the type being considered here is a function of cost and the rapidity of seasonal changes — every month, every two months, every quarter and every major season have been used. At the moment (1975) the operating cost of such a survey in East Africa is about US$50/1000 sq. km, not including salaries and the capital costs of equipment (aircraft etc.), minor equipment (recorders, cameras) and consumable items (films, recording tape).

Habitat data recording

Useful habitat data can be recorded from the air in any of these forms:

(1) Presence or absence in each of the sub-units on a transect, e.g. surface water, fires, flowing rivers, villages etc.

(2) On a scale of subjective estimates determined and agreed upon in the pre-survey reconnaissance phase. These can then be related to actual conditions by checking on the ground: e.g. a five-value greenness scale can be used to indicate the physiognomic stage of grass growth; similar five-value percentage scales can be used to express bush (i.e. woody dicotyledon) cover and grass cover. These subjective scale estimates can be very useful and consistent if made with care.

(3) Actual counts within the sample unit which allow estimates of the absolute numbers of features to be made in the same way as animal population estimates are derived, e.g. features such as termite mounds,

salt licks, water pans, villages, huts etc. Care should be taken to avoid confounding the observers' animal searching images by requiring them to record too many types of features in one survey. This can result in increased bias of the animal population estimates.

Examples of these forms of data recording are illustrated in Gwynne and Croze (1975).

Observers and observer variability

The best combination at present is an air crew of four per aircraft — a pilot, a front observer and two rear observers. Their exact duties will be discussed later. At this stage, however, it is important to consider the human element as most of the data will be collected by the observers on what is essentially nothing more than a subjective basis.

The greatest single source of error in current aerial census and monitoring surveys is the human observer (Savidge, 1973; Cobb, 1975). Individuals vary widely both in their ability to remain alert and fully functional when concentration is required for long periods, and in their ability to perceive objects as patterns with consistency. Few men will readily admit to being less efficient than their colleagues, where mental ability is concerned, and skill in counting animals from the air is no exception to this. So far comparatively little attention has been paid to this factor in monitoring programmes, particularly those which have been carried out by large units, such as game departments, where staff are freely and indiscriminately changed around.

Observer variability can be reduced by assessing their ability in a series of ground and aerial tests prior to the survey (see Watson, Freeman and Jolly, 1969; Watson, Jolly and Graham, 1969) — getting each to count quickly the animals in a series of colour slides shown on a screen is one such simple test. The score of each observer can then be related to the actual numbers of animals present and correction factors worked out. Further, from a number of pre-flight practice sessions it is possible to improve an observer's ability to estimate the numbers of animals in a group.

Some people are unsuited to this repetitive monotonous work due to variability in perception. Similar difficulties are encountered in other research fields such as microscope work for rumen content and blood smear analyses. Such people should not be used as observers in an aerial survey programme. Another cause of observer unsuitability is chronic airsickness.

Observer variability can be reduced by careful selection, prior to the survey, of the habitat data to be collected, with agreement on the definition of the character grades within each category: e.g. What does Greenness Category 5 really look like? What does 20 per cent bush cover look like? Much of this can be done with the aid of coloured photographic slides which can be used to train the observers.

Recent advances in the use of spectral reflectance data indicate that green biomass estimates may be collected by means of a photospectrometer mounted in the aircraft (Waddington, 1976). This, of course, eliminates observer error.

Further reduction in observer variability can be obtained by using, wherever possible, the same air crew on each repetitive survey. The data from successive surveys are, therefore, more comparable.

The problem of observer alertness can be reduced by ensuring that the flight times are not too long and that there is a short in-flight break between ending one transect and starting another. It is useful to ensure that the observers are recording at least one habitat parameter in addition to counting animals. This will ensure that they are obliged to concentrate and record data even over those regions which are devoid of animals; in other words recording activity helps combat boredom (Cobb, 1975).

It should not be thought that aerial surveys can only be carried out by trained biologists and that it is always necessary to have plane loads of graduates flying over the countryside. If the recording categories are well thought out and an observer training programme has been used there is no reason why junior staff cannot be observers, thus enabling the senior man to spend more time on data analysis. In a country where there is a shortage of trained biologists this approach may make the difference between being able or not being able to start a monitoring programme.

There are, however, drawbacks to using junior staff in this way and it is as well to be aware of them. The most serious is their possible lack of motivation stemming from disinterest in the project. This creates a tendency to regard observing as just another way of earning wages. The result may be continual lack of alertness and even invention of recorded data. Care in the selection and training of observers coupled with a system of checks and possibly special monetary inducement help eliminate this possible source of bias.

THE SYSTEMATIC RECONNAISSANCE FLIGHT (SRF)

The flight pattern

For most general census and monitoring purposes the best data gathering plan is to overfly the area on sampling flights using a systematic flight pattern: e.g. the aircraft could fly over the study area on parallel flight lines 10 km apart. These flight patterns can be repeated at different ecologically significant times (e.g. wet season, dry season).

Systematic flights allow data to be gathered in a form that permits quantification and at the same time shows how the animals and features are distributed spatially. Thus, for example, it is possible to obtain reasonable population estimates of the more abundant (and therefore important) large mammal species in the study area and to be able to follow their seasonal movements in response to changing ecological conditions, such as grass growth, availability of water etc.

Precise and accurate population estimates

A cause of confusion to those considering aerial sampling of animal populations for the first time is the indiscriminate use of the terms 'precise' and 'accurate' when talking of animal population numbers. By convention, in recent years, these have come to have distinct meanings (Caughley, 1972; Norton-Griffiths, 1973). A precise population estimate is one that has a small standard error but which may be divergent from the true numbers of animals present, that is, biased up or down. An accurate population estimate is one that is very close to the true numbers of animals, present (very little bias) but whose standard error is large. The ideal estimate is one that is both precise and accurate, but for both biological and sampling reasons this is rarely attained and the norm is an estimate that is less accurate and less precise than the ideal. The ultimate aims and funds available for any survey must govern whether the investigator strives for precision or accuracy — precision is best for detecting changes in population numbers with time whereas accuracy is best for considering large mammal biomass, primary production offtake by large mammals etc. (Caughley, 1974).

The SRF system of sampling normally provides less precise population estimates than other forms of aerial sampling, but the decrease in precision is more than offset by the value of the additional distribution data gained. If, however, the sampling transects are of sufficient length, even this defect is overcome producing population estimates whose confidence limits can be sufficiently small to be quite acceptable to the most discriminating population biologist.

Pre-SRF area familiarization

Before embarking on an SRF programme it is essential for the pilot and crew to familiarize themselves with the survey area and its major features. This should be done by:

(1) Study of available topographic maps.
(2) Study of available aerial photographs. If these can be assembled into print lay-down photo-mosaics, so much the better for they have more impact in this form. Study of satellite imagery is also worthwhile at this stage bearing in mind the very large scale which is used.
(3) One or more general reconnaissance flights over the area, flown at a much greater height above ground level than during an SRF. Altitudes varying from 300 m to 600 m have been found most suitable as these permit a broader overview than is possible at current SRF operational altitude (100 m).

SRF organization

Details of an SRF will vary depending on factors such as terrain, vegetation type, local seasons and the investigator's particular brief. Standardization of technique as much as possible is desirable, however, because it facilitates comparison between ecosystems — a typical SRF programme as practised in East Africa usually takes the following form.

Pre-flight planning

A map of the study area is overlaid with a reference grid such as the UTM 10×10 km grid system so that the grid, suitably numbered and lettered, can be used for orientation during flying and for presenting and analysing distribution data. This grid can often usefully be broken down into four smaller sub-units each of 5×5 km.

Systematic flight lines 10 km apart (e.g. centred on the grid system) are drawn on a 1:250 000 topographical map of the study area. A certain amount of latitude in placing these flight lines is permissable and advantage can be taken of this to locate commencement points in relation to prominent features such as river and road bends thus allowing the transects to be relocated on subsequent SRFs. Flight lines are normally orientated north–south or east–west as the shape of the area and wind conditions dictate. In general, however, it is best to avoid cross-wind flight lines (because of the problem of pilot navigation), very short flight lines (because of the problem of statistical treatment of data), and very long (over 100 km) flight lines (because of the problem of observer fatigue).

Flights should not be planned for very early or late hours because of the lack of light and the deep shadows which result. Flights should also not normally be planned for the midday period (10.00–15.00 hours), the time of maximum solar radiation during which many herbivores retreat into the shade for rest and rumination and are, therefore, not easily seen from the air. This behaviour pattern is modified by seasonal variation in cloud cover so that longer daily flight times can be considered for cloudy weather.

A decision should be made on the types of habitat data to be collected and the form in which these data are to be recorded bearing in mind the objectives of the SRF programme, the abilities of the observers and the terrain types to be overflown.

Aircraft operation

In East Africa most SRFs are flown by a crew of four in a single-engine, high wing aircraft; those most normally used are the Cessna series (180, 182, 185). Some workers, however, have used smaller two-seat, slower flying aircraft, such as the piper super-cub, but these are now normally used for special supplementary surveys such as the careful examination of hilly country for infrequent cryptic species such as greater kudu (*Tragelaphus strepsiceros* Pallas).

In a crew of four the pilot is responsible for air speed maintenance, height control, navigation and spatial location and for informing the crew of both transect numbers and subdivisions every 10 km flown. The front right observer, next to the pilot, documents habitat information (topography, drainage, vegetation type and cover etc., as well as greenness of vegetation, grazing intensity, presence of water, fire and other seasonal features). His duties may also involve recording the read-out from a digital spectrophotometer, if the aircraft is fitted with one, for green biomass estimates. In a series of SRFs made during a monitoring programme, the 'permanent' features need only be recorded once during an early flight.

The two rear seats are occupied by observers who mainly count animals by species between streamers or rods, two being fixed to the wing struts on either side of the aircraft (Savidge, 1976). The streamers are so spaced that at a particular height they subtend on the ground a strip of known width (e.g. 250 m) on either side of the aircraft. Thus along the flight lines the crew is counting animals (and features) in transects of known area. By proportionality, therefore, esticates of population size and variances can be made (Jolly, 1969). Animal species and numbers together with transect and subdivision co-ordinates are recorded by the observers using portable tape-recorders. Animal groups too large to count accurately (more than 20; e.g. herds of livestock) are photographed using a hand-held 35 mm camera. An automatic exposure, motorized, large magazine (250 picture) camera is ideal for this purpose though not essential.

Height control is critical since changes in altitude lead to fluctuations in strip width and thus to errors in counting animals — a rise causes the strip width to widen and a fall causes it to narrow. Height is best maintained by using a radar altimeter. In some circumstances, such as over flat terrain, an ordinary pressure altimeter can be used by relating it to contours and checking it periodically for diurnal pressure changes during the SRF by making low passes (2–3 m) over open ground where the altitude is known.

Navigation over flat featureless terrain is difficult. Aircraft can, however, be fitted with a Very Low Frequency (VLF) navigation system which permits transects to be flown with great accuracy. Repetitive flights along the same transect are possible, therefore, provided the starting point of each flight line can be relocated. The VLF navigation system is global in coverage so that it is only the aircraft instrumentation that is needed. This is expensive but its use is recommended, particularly over remote terrain where navigational errors could seriously jeopardize the results and their interpretation (Gwynne and Croze, 1975).

Data analysis

The data are all recorded with reference to a sub-unit on a light transect, that is within a particular grid square. On return from the flight the data are transcribed from the tape machines directly on to data sheets or computer coding sheets. At this stage, distribution of animals and the occurrence of habitat features such as greenness, water and vegetation type, may be plotted by hand on to gridded working maps of the area. The data may also be punched on to computer cards and transferred to magnetic tape for storage and analysis. Programmes are now in use which file the data and produce line-printed distribution maps as well as animal population and biomass estimates (Norton-Griffiths and Pennycuick, 1973; Cobb, 1975).

Computers save time and labour at this stage but they are not absolutely necessary to a monitoring programme. Indeed, many of the early monitoring programmes did not use them at all. Suitably gridded sketch maps for plotting distribution data, a portable electronic calculator with square root capability, and a knowledge of standard statistical techniques are all that is required. One of the new very small programmable electronic calculators with built-in statistical analysis capability would be ideal; the cost of these is now quite low and well within the budget of most projects. It must be emphasized again at this point, that the data collected should not exceed the analytical capabilities of the investigating team and should relate to the objectives of the programme. Failure to ensure this will lead not only to a waste of man-time and funds but will create a risk that nothing will be analysed and the programme will confuse rather than clarify land-use issues.

RESULTS

An SRF monitoring programme organized along the lines presented here and using very little in the way of sophisticated equipment ought to be able to produce at least:

(a) Estimates of the numbers and densities of the major herbivore species, both domestic and wild.

(b) A picture of the seasonal movements of the major herbivore species (this is dependent upon there being more than one SRF) leading ultimately to the development of probability of use maps.

(c) A broad scale vegetation map expressed in terms of plant physiognomy and/or plant cover.

(d) A broad-scale soil map expressed in terms of colour types.

(e) An outline of the areas that are important for wildlife which could be used to delimit, if required, possible reserves and parks.

(f) An outline of the areas that are important to livestock. Such information could be used for the planned control of stock numbers and in rangeland development programmes.

(g) A land-use map(s) in terms of human occupance, showing activities in rangeland, agriculture, forestry, etc.

(h) If the SRF programme is continued over several seasons, a delineation of productive and non-productive areas.

All these are types of information that are important to land and resource management and have real practical value to the development planners. The results are, however, all base-line descriptions. For causal relationships and for the details of the interactions we see from the air, we must eventually get back to earth. We begin then to think of initiating a full-scale scientific ecological monitoring programme which will include all the necessary ground investigations. And, at the far end of the scale of ecological management ambitions, we must begin to delve deeper into the possibilities that the sophisticated satellite sensors offer (Gwynne and Croze, 1975).

REFERENCES

CAUGHLEY, G. (1972) Aerial survey techniques appropriate to estimating cropping quotas. (*KEN:SF/FAO 26*, Work Paper, No. 2), pp. 1–13.

CAUGHLEY, G. (1974) Bias in aerial survey. *Journal of Wildlife Management*, Vol. 38, No. 4, pp. 921–33.

COBB, S. (1975) Preliminary results of the aerial monitoring programme in the Tsavo region. A report to the trustees of the Kenya National Parks. (Animal Ecology Research Group, University of Oxford). Pp. 23.

GWYNNE, M. D. and CROZE, H. (1975) East African habitat monitoring practice: a review of methods and application. (Proceedings of the International Livestock Centre for Africa Symposium on *Methods of surveying and mapping rangelands*, Bamako, Mali).

JOLLY, G. M. (1969) Sampling methods for aerial censuses of wildlife populations. *East African Agricultural and Forestry Journal*, Vol. 34, pp. 46–9.

NORTON-GRIFFITHS, M. (1972) *Serengeti ecological monitoring program*. (African Wildlife Leadership Foundation, Nairobi).

NORTON-GRIFFITHS, M. (1973) Counting the Serengeti migratory wildebeest using two-stage sampling. *East African Wildlife Journal*, Vol. 11, pp. 135–49.

NORTON-GRIFFITHS, M. and PENNYCUICK, LINDA (1974) Trend surface analysis (MX 23). Tech. Pap. Inst. Dev. Stud., Nairobi). Pp. 1–10.

PENNYCUICK, LINDA (1975) Movements of the migratory wildebeest population in the Serengeti area between 1960 and 1973. *East African Wildlife Journal*, Vol. 13, pp. 65–87.

SAVIDGE, J. M. (1973) Aerial census techniques in estimating wildlife populations. (UNDP/FAO Kenya Wildlife Management Project, Nairobi).

SAVIDGE, J. M. (1976) Fixed strut-rods as an aid to the aerial counting of large mammals.

WATSON, R. M. (1969) The planning of flights and the handling of time-serial data. *East African Agricultural and Forestry Journal*, Vol. 34, pp. 70–8.

WATSON, R. M., FREEMAN, G. H. and JOLLY, G. M. (1969) Some indoor experiments to simulate problems in aerial censusing. *East African Agricultural and Forestry Journal*, Vol. 34, pp. 56–9.

WATSON, R. M., JOLLY, G. M. and GRAHAM, A. D. (1969) Two experimental censuses. *East African Agricultural and Wildlife Journal*, Vol. 34, pp. 60–2.

3

LARGE SCALE MEASUREMENT OF HABITAT STRUCTURE AND CONDITION, AND ITS USE IN INTERPRETING ANIMAL DISTRIBUTION

Stephen Cobb

Animal Ecology Research Group, Department of Zoology, University of Oxford, UK

INTRODUCTION

Attention has been drawn to some of the advantages that appear to attach to the collection of data on large mammals in a large-scale grid format. In this paper, this belief is taken a little further by describing some of the ways in which the habitat in Tsavo was described, and how this can be used as a framework for interpreting the animal distribution data.

First of all, it should be stressed that the resolution of what was done was coarse; the results are not comparable in their accuracy to those obtained by more conventional methods of aerial photographic interpretation, vegetation mapping, soil survey and land classification. There are two reasons why this is not considered disturbing. Firstly, the area of study was large, (around 45 000 sq. km) and the manpower necessary to conduct the sort of surveys mentioned was not available (although vegetation mapping and soil survey are now under way in Tsavo) as had been the fortunate case in the Serengeti (e.g. Gerresheim, 1974; de Wit, 1973; Herlocker, 1974). In addition, the aerial photo-cover of Tsavo, although complete, was found (Norton-Griffiths, 1973) to be of inadequate quality to fulfil the requirements of this type of survey. Secondly, the sort of information that was required, namely that for interpreting animal distribution, could for the most part be collected as a part of animal counting exercises; certain information had, of necessity, to be collected as a separate exercise, in particular climatic data, while other aspects of the system, such as the soils and topography, could always be added to the store of information as they became available.

DATA COLLECTION

Climate

The location of Tsavo is rather fortunate in this respect, in that, despite much of it being very remote, it is next to several densely populated areas and is bisected by the Uganda Railway, along which climatic data has been gathered for the last 70 years. There were 12 meteorological stations either inside the 45 000 sq. km study area, or close enough to it to be of interpretative value. In addition, data were available from over 100 rain-gauges in and near the area, though some of the data were incomplete for the period of our programme; also, the network of gauges in the park was improved and rationalized — there are now 56 gauges within the park, either being read daily at gates and at ranger posts, or storage gauges are used, which have oil on the water surface, and are read monthly by research staff.

The result of all this rainfall information has been that we have been able to describe the probable rainfall total for each of 450 grid squares for each month. This has been done in the first instance by feeding all the available data into a computer program, available at Oxford University, that generates isohyets. This shows the total rainfall for the climatic year and gives a very satisfactory picture of how the climate varies, spatially, across the Tsavo region. By superimposing a grid over the isohyets of the monthly totals of the month (or two months) preceding each flight, actual values for each square are available for analysis alongside the other environmental information, whose collection is described below. Similar transfer techniques are used for the other climatic data.

Habitat

On each of the eight survey flights conducted over the Tsavo region, information about the gross structure and phenological condition of the vegetation was collected; attempts were made to quantify other aspects of habitat as well. This task was primarily the job of the front right-hand observer in the aircraft. Because of overloading the observers, information could not be collected on all variables on all flights. Anything certain to change from census to census, such as water availability or the greenness of the grass, was quantified on each occasion. Constants, such as vegetation type or soil type, were each quantified on more than one occasion, so that the validity of assessments could be checked. A list of all the attributes quantified is set out below.

Constants

(1) Vegetation type: a purely structural classification, based on the rangeland classification system of Pratt et al. (1966).
(2) Dominant tree species: the three most numerous tree species, or in their absence, bushes.
(3) Woodland damage: degree of damage to trees, in 20 per cent categories; largely attributable to elephants, though locally fire and illegal charcoal burning (outside the park) were important.

(4) Soil type: based largely on colour, to a lesser extent on texture (particularly in areas of lava and of young river-dissection, stony soils predominate).
(5) Canopy cover: measured in 20 per cent categories.
(6) Bush cover: measured in 20 per cent categories.
(7) Grass cover: measured in 20 per cent categories.
(8) Grass height: assessed approximately in 10 cm height intervals.

Variables

(a) Grass condition: assessed as an interval scale of water content, from very green to very dry.
(b) Browse condition: measured in the same way as the grass condition, though clearly with rather more errors inherent in it.
(c) Water availability: this not only differentiated (i) between artificial sources of water (boreholes and Haffir tanks, for example, exist outside the park) and permanent natural ones, i.e. springs; and (ii) between rivers and rainwater pools; but noted if there was complete absence of water (the most normal condition).

With this information and that on the total annual rainfall, the climate and the rainfall of the preceding months, an extensive body of information for interpreting animal distribution was available.

Calibration

Flying over the bush fast and low, trying to assess a whole range of things about the habitat, was one thing; finding out whether it represented the true situation on the ground was another. Fortunately, research was being done on the ground at the same time and this has made it possible to test the validity of our measurements in certain test areas, mostly in the centre of the park. The constants have been the easiest to calibrate. For example, there is low-level aerial photography, kindly performed by Dr Norton-Griffiths (1973), for the tree damage. The work on tree regeneration currently being performed by Mr T. F. Corfield includes data, largely collected by the Point Centre Quarter method, for calibrating all the cover categories and the tree species. Regular measurements of grass height in a range of grassland communities were made by the author. The recent arrival of a Dutch soil scientist means that the soil type classification can also be calibrated. The data on the grass and browse condition are less easy to calibrate, but tests comparing this sort of data with absolute measures of greenness, obtained from infra-red satellite photographs, done in neighbouring Amboseli by Dr David Western, indicate that agreement is very close. It is therefore possible to estimate how satisfactory the study reported has been.

Data Analysis

The first thing that could be done with the habitat data was a simple mapping exercise. In the first case woodland damage can be shown. A map demonstrates the nature and extent of the damage, heaviest within the park and close to permanent water supplies. The state of affairs may have been intuitively obvious, but it does provide an objective picture that can be used in management planning: some parts of the park are in a very different state to others. The second case, the change from wet to dry season in a variant, grass greenness can be demonstrated. The rainfall patterns for the two months preceding these two counts, i.e. April rain for the May count and August rain for the dry season count in September. Again, these provide both a permanent record of the state of things at a fixed moment, and a picture (of which the park management should be aware) of absolute differences in the way in which the landscape varies with season from one end of the park to the other.

Thereafter the analysis becomes more complicated. One of the things that was organized early in the research was the preparation of a computer program, large and complex enough to handle a great mass of data, using the sort of computer that is typically available in Africa. Apart from the production of simple maps of the habitat data, the program produces population estimates for each species, to which reference was made elsewhere. The animal numbers are then turned into densities of each species (per sq. km), that can also be output in the form of maps. The densities of animals can be produced in exactly the same serial format as the habitat data. A second program has been written which combines all the information for each point on the map on one line, whether it is animal data, habitat data, or rainfall data collected after the event does not matter. This was done via two separate computer programs, because of the limitations on the size of the computers available in East Africa.

It is not the purpose of this paper to report on the actual results of this research. However, it is relevant to give an idea of the sort of results that can be obtained. This has been done using two computer programs, so-called packages of statistical analyses. These incorporate a whole range of multivariate analyses, which enable really very complicated questions to be asked. For example, consider an early dry-season map of elephant distribution and then a map of grass greenness for the same census; the elephants appear to be where the grass has remained green longest — not very surprising. Information may also be available for other factors controlling their distribution which are as important, but not so obvious. The multivariate tests show just how important each one is. They measure the relative importance of each variable by showing how much of the variation in the dependent variable, in this case elephant density, is explained by each of the independent variables. The results might indicate for example that grass greenness explained 40 per cent of the variation, while soil type, water availability and canopy cover each explained only 10 per cent of the variation. This means that grass greenness is, as suspected, the most important factor, at the time of year, controlling elephant distribution; the vital piece of knowledge is that it is now known how much more important is grass greenness than canopy cover. Changes in canopy cover (for example, due to elephant damage) are concluded to be far less likely to affect the elephant distribution than changes in the rainfall. This gives the research not only descriptive power, but also predictive power; prediction being the meat on which the wildlife manager feeds. This sort of analysis would seem to be most appropriate for solving management problems.

DISCUSSION

Measuring vegetation without actually touching it may seem to botanists to be an absurd exercise. There are certainly inherent limitations, but if these are understood and borne in mind, it does seem that this is an economical and promising way of gathering data for interpreting animal distributions. There are two potential drawbacks which have not yet been mentioned and which will be discussed briefly here.

The first concerns the quality of the data and the variables that have been measured. There is no way that multivariate analyses are going to relate animal distribution to variables that have not been measured. For example, sodium and calcium levels have been shown elsewhere (Weir, 1972; Kreulen, 1974) to be extremely important determinants of distribution. If it was considered important that mineral content of the soil in analysis be included in studies involving flying exercises, then it would be necessary to assess this through ground control work. This is often impractical but was done for as many grid squares as possible, and then add it to our list of constants. Data from the present study in fact shows that wart-hog distribution appears to be strongly correlated with soil type, which is not a surprising fact for a burrowing animal. Although, unfortunately, there is no means of knowing whether it is the soil colour alone, or the soil texture, or its chemical composition that matters most to the wart-hog, the data does indicate where to look next. That next look will, in most cases be from the ground, if the need for a more detailed answer is great enough to justify the extra cost. The important thing to remember is that, had soil type explained, say, 50 per cent of the variation in wart-hog distribution, the fact that the physical and chemical properties of the soil had not been measured would not alter this; that 50 per cent would simply be masking the variation attributable to the constituents of the soil type, each of which is strongly correlated with the soil type itself. Only when four, or perhaps five variables between them fail to explain a total of 80 per cent of the variation in a species distribution, can one say that one has been measuring the wrong things.

The second potential drawback concerns the fact that the nature of the Tsavo system changes very markedly from one end to the other. The rainfall has been demonstrated to change, in a typical year, from 1200 mm to 200 mm in just a few tens of kilometres. It is, therefore, likely that the factors that control, say, zebra distribution in moist Tsavo West are not the same as those which control their distribution in the arid north of Tsavo East. To offset this drawback, exactly the same data can be used to define distinct regions (if they exist) by a hierarchical process of several different multivariate analyses. This type of approach has already been attempted briefly for Ruaha National Park, Tanzania, by Norton-Griffiths (1975). Once the regions have been defined, they can be used to ask questions of the form: 'Does zebra distribution vary in Area A for the same reasons as it does in Area B?' Since in large wildlife areas, or large rangeland areas generally, different management practices should be executed in different parts of the area, this ability to differentiate is clearly of considerable use to management.

To conclude, then, it must be borne in mind that data collected in the way described, needs to be interpreted (and collected in the first place) with some care. If proper control is exercised, however, the data can be extremely informative, both in their own right and as the key to the interpretation of animal distribution data. It is not proposed that this methodology should replace detailed habitat studies on the ground. It is, nevertheless, believed that on account of the relative ease of execution and the power of the associated analytical techniques, the method could be most valuable in at least four types of situation that might arise in this part of Africa:

(1) As a preliminary survey in areas where no previous research has been done, even if these are small.

(2) In large areas, where detailed studies on the ground cannot be extended to account for the whole area — this may be of great value if the detailed studies are performed (as was the case in Tsavo) in a typical area.

(3) In areas where access is difficult or the vegetation is unsuitable for groundwork (both these were true for parts of Tsavo).

(4) Particularly in semi-arid areas, where the vegetation is more open; this type of area may well be dominated by pastoralists and their livestock, towards a comprehension of whose dynamics the methods seem very well suited.

That each of these situations has management problems, many of which could be better understood in the light of the results that this type of research would generate, barely needs repeating.

ACKNOWLEDGEMENTS

The research described in this paper has involved many people, both in Kenya and in England. More than to anyone else, I owe a debt of gratitude to Dr David Western for endless fruitful discussion and close collaboration. Dr Walter Leuthold and Mr Tim Morgan, of the Tsavo Research Project, put in countless hours of work, both inside the aircraft and out; without their efforts and those of the other members of T.R.P., nothing would have been achieved. Dr Mike Norton-Griffiths has been a fund of useful comment on data analysis. Mr Andrew Dearing wrote the computer program. Others, too numerous to mention, have assisted in many ways.

Finally I would like to thank the Director and Trustees of Kenya National Parks for permission to do the research, and the Wardens of Tsavo East and

Tsavo West National Parks for their practical assistance and enthusiasm for the project.

To all these people I am exceedingly grateful.

REFERENCES

GERRESHEIM, K. (1974) *The Serengeti Landscape Classification.* (A.W.L.F., Nairobi).

HERLOCKER, D. J. (1974) Woodland Mapping. *S.R.I. Annual Report 1973–1974*, pp. 20–1.

KREULEN, D. (1974) Feeding ecology and movements of migratory Wildebeest. *S.R.I. Annual Report 1973–1974*, pp. 47–9.

NORTON-GRIFFITHS, M. (1973) *Report on the Aerial Photography carried out in the Tsavo National Parks.* (A.W.L.F., Nairobi).

NORTON-GRIFFITHS, M. (1975) The numbers and distribution of large mammals in Ruaha National Park, Tanzania. *East African Wildlife Journal*, Vol. 13, pp. 121–40

PRATT, D. J., GREENWAY, P. J. and GWYNNE, M. D. (1966) A classification of East African rangeland, with an appendix on terminology. *Journal of Applied Ecology*, Vol. 3, pp. 369–82.

WEIR, J. S. (1972) Spatial distribution of elephants in an African National Park in relation to environmental sodium. *Oikos*, Vol. 23, pp. 1–13.

DE WIT, H. (1973) Soil Studies. (*S.R.I. Annual Report 1972–1973*), pp. 36–7.

4

MATRIX APPROACH TO POPULATION DYNAMICS

J. K. Egunjobi

Department of Agricultural Biology, University of Ibadan, Nigeria

INTRODUCTION

Population dynamics is one aspect of ecology which has a strongly mathematical bias. Matrix algebra, for example, is often used. A matrix is a rectangular or square array of numbers arranged in rows and columns. For example

$$
\begin{bmatrix}
a_{11} & a_{12} & a_{13} & a_{14} \\
a_{21} & a_{22} & a_{23} & a_{24} \\
a_{31} & a_{32} & a_{33} & a_{34} \\
a_{41} & a_{42} & a_{43} & a_{44}
\end{bmatrix}
$$

is a 4×4 square matrix. Each element of the matrix is referred to Aij where i represents the row and j the column. A matrix consisting of a single column ($1 \times n$ matrix) is called a column vector. Thus

$$
\begin{bmatrix}
a_{11} \\
a_{21} \\
a_{31}
\end{bmatrix}
$$

is a vector. Conversely, a matrix that is just a row is a row vector. In the matrix nomenclature a single number is referred to as a scalar.

The matrix application to population studies was originally described by Lewis (1942) and then independently by Leslie (1948, 1959). The model has since been modified for various applications by Lefkovitch (1965), Pennycuick *et al.* (1968) and Williamson (1959, 1967). Lately, Usher (1966, 1967, 1969a, 1969b) has extended the use of the model into the study of plant populations and forest management. In a further paper Usher (1972) described an adaptation of the model which is suitable for studying the cycling of nutrients and the flow of energy in an ecosystem.

Biologists and wildlife managers are generally not mathematically oriented. The main object of this paper is to interest wildlife biologists in the relatively easy methods of matrix manipulation for studies of animal populations. The paper is divided into four parts. The first part deals with the structure of the basic model, and its biologically meaningful mathematical properties. In the second part a hypothetical population of rats is used to illustrate the use of the model. In the third, a comparison of the model with that of calculus is made and the fourth focuses attention on the difficulties of applying the model to wildlife studies.

THE BASIC LESLIE MODEL

The basic Leslie matrix model can be written in matrix notation as

$$A = \begin{bmatrix} f_0 & f_1 & f_2 & f_n \\ P_0 & 0 & 0 & 0 \\ 0 & P_1 & 0 & 0 \\ 0 & 0 & P_0 & 0 \end{bmatrix}$$

in which the columns of the matrix represents convenient age groups, the elements $f_0, f_1 \ldots .f_n$ in the first row represent the age specific fecundity terms, i.e. the number of reproducing females that each individual in each age group can produce between time t_i and t_i+1. The sub-diagonal elements P_0, P_1, P_2 represent the probability that a female animal age i at time t_i will survive till time t_i+1.

In short, the matrix A is described as the survival fecundity matrix.

$at+1$ is another column vector, similar to at representing the new age structure after one time interval. It is obtained by simply pre-multiplying the column vector at with the matrix A.

Thus we have:

$$\begin{bmatrix} f_0 & f_1 & f_2 \ldots \ldots f_n \\ P_0 & 0 & 0 \ldots \ldots 0 \\ 0 & P_1 & 0 \ldots \ldots 0 \\ 0 & 0 & P_n \ldots \ldots 0 \end{bmatrix} \begin{bmatrix} at,0 \\ at,1 \\ at,2 \\ at,n \end{bmatrix} = \begin{bmatrix} at+1,0 \\ at+1,1 \\ at+1,2 \\ at+1,n \end{bmatrix}$$

$$Aat = at+1.$$

A repeated application of this matrix predicts the age structure at specific time intervals.

The survival-fecundity matrix A is a square matrix, and its elements are either f, p or zero; and since neither of these can take a negative form ($f \geqslant 0$, $0 \leqslant P \leqslant 1$) A is described as non-negative matrix. The mathematical properties of such a matrix have been investigated by Leslie (1945, 1948), Sykes (1969) and Searle (1966). These treatments are beyond the scope of this exposition. For a biological application of these properties see Searle (1966) and Usher (1972). The most biologically relevant of these properties are:

(1) For an $n \times n$ square matrix, there are n latent roots (eigen values) and vectors (eigen vectors) which satisfy the equation.

$$Aa = \lambda a \qquad (2)$$

where λ is any latent root and a is a latent vector associated with λ.

For an understanding of the theory of eigen values and vectors of a matrix the reader is advised to read Chapters 3 and 7 of Searle's *Matrix algebra for biologists*.

(2) One of the latent roots of the matrix is dominant.

(3) Corresponding to this latent root is a latent vector, having all its elements non-negative, thereby giving a biologically meaningful population structure.

(4) The natural log of the dominant latent root is the intrinsic rate of natural increase, i.e. there exists a constant λ such that $At+1 = \lambda\, at$.

Using this type of matrix operation, a steady state or stable age distribution can be estimated.

APPLICATIONS OF THE BASIC MODEL

To illustrate the operations of the model a captive population of cane rat (*Thryonomys* spp.) is used. Suppose each female rat produces its first litter of 8 when it is 12 to 15 months old, thereafter producing the same size of litter a year later and reducing the litter size to 6 after another year before dying. Suppose the probability of a young animal surviving to the second and third years is 0.5 in each case, but increases to 0.8 between the third and fourth year. If we assume that the sex ratio at birth is 1 :1   and that this is constant, the survival–fecundity matrix is:

$$A = \begin{bmatrix} 0 & 4 & 4 & 3 \\ 0.5 & 0 & 0 & 0 \\ 0 & 0.5 & 0 & 0 \\ 0 & 0 & 0.8 & 0 \end{bmatrix}$$

(Note that the young animals age <1 year are not fecund.)

Starting with an initial population of:

$$\begin{bmatrix} 4 & 6 & 1 & 0 \end{bmatrix}$$

The number and age structure of the population after 1, 2, 3, n years can be predicted. By pre-multiplying the vector with the matrix A, the model predicts the population to be

$$\begin{bmatrix} 28 & 2 & 3 & 1 \end{bmatrix}$$

after 1 period of time and

$$\begin{bmatrix} 23 & 14 & 1 & 2 \end{bmatrix}$$

after 2 periods of time.

It should be noted that in each case the vector gives the age distribution of survivors and offspring of the original population.

With this approach one may wish to experiment in the model by varying the probability terms of the matrix.

The latent roots of the matrix can be solved from the characteristic equation:

$$\begin{bmatrix} 0-\lambda & 4 & 4 & 3 \\ 0.5 & 0-\lambda & 0 & 0 \\ 0 & 0.5 & 0-\lambda & 0 \\ 0 & 0 & 0.8 & 0-\lambda \end{bmatrix} = 0$$

By diagonal expansion the value λ can be solved. In a 2×2 matrix, a manual algebraic expansion of the characteristic equation is enough for solving for λ. With larger matrices, this becomes very cumbersome for non-mathematicians. Computer programs are now available for finding the latent roots of large matrices.

The original Leslie matrix model was designed for the female population. However, Williamson (1959) has developed the model to include both sexes.

A COMPARISON WITH THE CALCULUS MODEL

Most biologists are familiar with the exponential growth model:
$$N_t = N_o \exp(rt) \tag{3}$$
for a single population when no limitations are placed upon the growth in numbers, and the more realistic Verhulst–Pearl (logistic) model:
$$N_t = \frac{k}{1+0 \exp(rt)} \tag{4}$$
in which the effect of increasing population is taken into consideration. These models suffer from a major defect arising from the fact that the age structure of the population and the differences of birth and death rates between the different age groups are not considered. In these models the population is taken as if it were a cohort without differences in these vital statistics. This major defect is overcome in the matrix model, in which the population is broken into age groups with specific fecundity and mortality rates. On the other hand, however, the matrix model shares some of the defects of the exponential growth model in that it does not consider the effects of increasing population on the fecundity and mortality terms of the matrix. With increasing population, resources are in short supply and competition between individuals reduces their reproductive rate or their survival. The predictions based on the basic matrix model will be close to reality over a small number of generations until

the effect of increasing population starts to tell on the fecundity and mortality terms. Noting this defect in the basic matrix, Leslie (1948, 1959), Williamson (1959) and Pennycuick *et al.* (1968) investigated various ways in which the effect of increasing density can be introduced into the terms of the basic matrix model. In the modification by Pennycuick *et al.* (1968) the fecundity and survival terms in the matrix were made to depend on the total number of animals in the population, so that the p and f values do not remain constant but change value after each iteration. The fecundity and survival terms decreased with increase in the population. With this modification the major defect to the use of the matrix model can be removed. The matrix model has obvious advantages over the well-known calculus models. Usher (1972) listed four such advantages.

One obvious advantage of the matrix model is the easier mathematics of computation. Integral calculus is a more difficult concept for biologists to grasp than matrix algebra. In addition, the calculus model implies that growth is a continuous process, taking place in infinitesimally small intervals, whereas the growth of animal populations takes place in discrete units of time. Events that occur in discrete units of time are better modelled with matrices which also operate in discrete units. Furthermore, matrices are more easily handled by various computer languages, especially the algebraic-like Fortran.

THE USE OF MATRIX MODELS IN WILDLIFE STUDIES

Population studies of animals in the wild are compounded by factors such as predation, immigration, emigration and stochastic elements of the environment. Although the survival–fecundity matrix describes the transition probability from one time to another the model is still essentially deterministic, and not stochastic. For an efficient prediction of populations it is necessary to incorporate stochasticity into the model. Pollard (1966) developed a stochastic form of the basic model for predicting human populations in Australia. While it may be assumed that a factor such as predation may be incorporated into the survival terms $(P_0, P_1 \ldots P_n)$, as has been done by Pennycuick *et al.* (1968), immigration and emigration cannot be easily quantified and incorporated into the model. These are events which are controlled from outside the population.

A major prerequisite to matrix modelling is the preparation of a life table. Except in some well-known cases, the ageing of animals living in the wild is extremely difficult. Data on such attributes as fecundity and survival have to be obtained from captive animals, which invariably behave differently under laboratory conditions and different diets.

Despite these limitations, reasonable predictions with matrix models have been obtained (see for example, Usher (1972) on blue whale *(Balaenoptera musculus)* and on red deer *(Cervus elaphus)*.

Matrix modelling offers a considerable advantage to management for simulation of various harvesting regimes. Adaptation of the basic model for investigations of the effect of exploitation has been made by Lefkovitch (1967) and Williamson (1967); Williamson (1967) has shown that the value of the dominant eigen value λ can be used to estimate what proportion of the animal population can be harvested. As mentioned previously in this paper one of the important attributes of λ is that $\log_e \lambda = r$ which is the intrinsic rate of growth. Williamson (1967) has shown that if the population size increases from N_0 to N_i over a period of time then the percentage of the population that can be taken is:

$$H = 100\,(\lambda - 1)/\lambda$$

A matrix model can be used in this way to estimate what fraction of the population can be cropped at time intervals.

Another consideration is how best to distribute the harvest within the population and yet retain a stable age distribution. This aspect has also been investigated by Lefkovitch (1967) and Williamson (1967). By performing experiments on the model (computer gaming) one may arrive at a figure that gives the maximum harvesting and still retains a stable age distribution.

Matrix models can also be used in predicting populations of domesticated wildlife. Two recent studies, Ajayi (1974) and Asibey (1975) have shown that some species of wildlife can be domesticated and kept by families to supplement their source of animal protein. The use of a model of this type can be of considerable importance for advising families on how many animals can be profitably kept, and what proportion of the population should be cropped.

REFERENCES

AJAYI, S. S. (1974) The biology and domestication of the African giant rat *Cricetomys gambianus* Waterhouse. (Ph.D. thesis, University of Ibadan).

ASIBEY, E. O. (1975) Some ecological and economic aspects of the grasscutter *Thryonomis swinder; anus* Temminck Hystricomorpha Rondentia mammalia in Ghana. (Ph.D. thesis, University of Aberdeen).

LEFKOVITCH, L. P. (1965) The study of population growths in organisms grouped by stages. *Biometrics*, Vol. 21, pp. 1–18.

LEFKOVITCH, L. P. (1967) A theoretical evaluation of population growth after removing individuals from some age group. *Bulletin of Entomological Research*, Vol. 57, pp. 437–45.

LESLIE, P. H. (1945) On the use of matrices in certain population mathematics. *Biometrika*, Vol. 33, pp. 183–212.

LESLIE, P. H. (1948) Some further notes on the use of matrices in population mathematics. *Biometrika*, Vol. 35, pp. 213–45.

LESLIE, P. H. (1959) The properties of a certain lag type of population growth and the influence of an external random factor on a number of such populations. *Physiological Zoology*, Vol. 32, pp. 151–9.

LEWIS, E. G. (1942) On the generation and growth of a population. *Sankhya, Indian Journal of Statistics*, Vol. 6, pp. 93–6.

PENNYCUICK, C. J., COMPTON, R. M. and BECKINGHAM, I. (1968) A computer model for simulating the growth of a population or of two interacting populations. *Journal of Theoretical Biology*, Vol. 22, pp. 381–400.

SEARLE, S. R. (1966) *Matrix algebra for the biological sciences (including application in statistics)*. (John Wiley & Sons). Pp. 296.

SYKES, Z. M. (1969) On discrete stable population theory. *Biometrics*, Vol. 25, pp. 285–93.

USHER, M. B. (1966) A matrix approach to the management of renewable resources, with special reference to selection forests. *Journal of Applied Ecology*, Vol. 3, pp. 355–67.

USHER, M. B. (1967) A structure for selection forest. *Sylva*, Vol. 47, pp. 6–8.

USHER, M. B. (1969a) A matrix model for forest management. *Biometrics,* Vol. 25, 309–15.

USHER, M. B. (1969b) A matrix approach to the management of renewable resources, with special reference to selection forest — two extensions. *Journal of Applied Ecology*, Vol. 6, pp. 347–8.

USHER, M. B. (1972) Developments in the Leslie Matrix Model. In *Mathematical models in ecology.* Jeffers, J. N. R. (Ed.) 12th Symposium of the British Ecological Society. (Blackwell Scientific Publications, London). Pp. 29–60.

USHER, M. B. (1973) *Biological conservation and management.* (Chapman & Hall). Pp. 394.

WILLIAMSON, M. H. (1959) Some extensions of the use of matrices in population theory. *Bulletin of Mathematical Biophysics*, Vol. 21, pp. 13–17.

WILLIAMSON, M. H. (1967) Introducing students to the concepts of population dynamics. In *The Teaching of ecology.* Lambert, J. M. (Ed.) (Blackwell, Oxford). Pp. 169–75.

5

A MODEL FOR PREDICTING AGE IN UNGULATES

C. A. *Spinage*

College of African Wildlife Management, Mweka, Moshi, Tanzania

INTRODUCTION

The development of wildlife management in Africa has led to a need for more precise assessments of the ages of animal population members. At one time it was considered sufficient to lump the members into arbitrary categories of wear, such as young, middle-aged, old and very old: often denoted as A, B, C and D. This tells us little about the population under scrutiny. It cannot, for example, tell us the turnover rate. Some of the advantages of more precise age determination have been outlined by Spinage (1973); but the objective of the study must always be born in mind, for too fine precision becomes more of academic interest than of practical use. A few years ago, however, we had little idea of the expectation of life of most African mammals, and we are still mainly dependent on Flower's 1931 work for our knowledge of the lifespans of many of them.

To determine precise age without a knowledge of the date of birth, appears to be virtually impossible in tropical African ungulates. The nearest approach is obtained from counts of the rhythmic cementum lines laid down in the root of a tooth; but even these are variable in occurrence, either as a result of faulty preparatory technique, or from physiological phenomena: to the extent that even they can only provide an approximate age (Spinage, 1975). Their study, furthermore, necessitates skilled technique, specialized laboratory equipment, and time: requirements which are often unavailable to the field worker. Based on principles of tribology, Spinage (1971) therefore proposed a model for determining specific age from the first molar crown height.

THE WEAR MODEL

The model incorporates three assumptions; that individual variations in the rate of wear of a tooth are normally distributed, that the crown of the first molar is worn to zero at the termination of the physiological lifespan of the species, and that the first molar would have an initial height in the absence of wear. The first assumption is in accord with basic biological principles. The second is

33

based on the fact that many ungulates which die of old age, often do have their first molar worn to the roots, and that evolution tends to select for maximum economy in the use of resources; generally no animal is provided with structures which are capable of outlasting its physiological lifespan by a significant amount. The final assumption is necessary as the tooth begins to wear as soon as it cuts the gingiva, although full crown height may not yet be attained. Accepting these assumptions, the model requires a knowledge of only two parameters: the lifespan of the species, and the initial height of the first molar tooth — before wear. The latter will be close to the maximum height reached after eruption. Using these two parameters it was found by experimentation, that the rate of wear of the first molar from emergence to extinction, most closely approximated to a square root function, of the form

$$Y = Y_0 \left(1 - \frac{t}{n} \right)^K$$

where Y is the molar crown height at a given time (age) t, Y_0 is the height of the tooth in the absence of wear, n is the lifespan of the species, and k is a constant: in this case, the square root. Given that wear follows this pattern, we can predict, on average, the age of a molar tooth from its height alone.

Variations in the rate of tooth wear

In some individuals the teeth will wear more quickly than in others, in some they will wear more slowly; but on average, they will wear at a rate shown by the model. Chaplin (1971) postulated however that wild animals should show more variable rates of tooth wear than that shown by domestic animals, although the latter was variable enough already. As I have suggested elsewhere (Spinage, 1973), the opposite should be the case, as domestic animals are not selected for the hardness of their teeth. The feeding habits of wild populations tend to be uniform and one might expect that evolution will have selected those most adapted to the environment. The studies of Steenkamp (1974) on indigenous and exotic cattle in Rhodesia support this view. Such inaccuracies that variable rates of wear might induce in the model, will be no different to those which they would induce into estimation of age based on the visual appraisal of wear.

Sexual dimorphism

Before using the model for both sexes it is necessary to ascertain whether sexual dimorphism in tooth size exists. This was not apparent in the defassa water-buck *Kobus defassa* Ruppell (Spinage, 1967), in the impala *Aepyceros melampus* Lichtenstein (Spinage, 1971) and in the buffalo *Syncerus caffer* Sparrman (Grimsdell, 1973). A small sample of ten male and ten female Grant's gazelle *Gazella granti* Brooke, was compared in detail; the sample being of equal arbitrary age groups based on visual appraisal, according to the

TABLE 1

Comparison of the weight (g) and length (mm) of the maxillary tooth row of male and female Grant's gazelle

	Mean n	Mean weight	S.D.	Range	Mean n	Mean length	S.D.	Range
Male	10	22.9	8.49	22–30	10	80	8.17	75.5–82.5
Female	8	18.5	7.72	18–26	10	77.1	1.73	74–81
t-test		1.137				3.35		
P		0.3 0.2				0.01–0.001		

S.D. = Standard deviation.
P   = probability that the populations do not differ.
n   = number of individuals in the sample taken.

TABLE 2

Comparison of the dimensions and weight of the maxillary first molar of the male and female Grant's gazelle (in mm and g)

	n	Mean crown height	S.D.	Mean weight width	S.D.	Mean crown length	S.D.	Mean crown length	S.D.
Male	10	10.6	2.46	3.97	0.62	11.65	0.29	14.35	0.67
Female	10	11.35	2.32	3.79	0.75	11.53	0.80	14.45	0.73
t-test		0.701		0.587		0.448		0.321	
P		0.5		0.6 0.5		0.7 0.6		0.8 0.7	

S.D. = Standard deviation.
P   = probability that the populations do not differ.
n   = number of individuals in the sample taken.

TABLE 3

Comparison of weight (g) and dimensions (mm) of the maxillary first molar of the male and female Grant's gazelle, for equal crown heights

Crown height	Weight (g)	Male Crown length	Crown width	Weight (g)	Female Crown length	Crown width
13	4.69	14.5	11.5	3.58	14	11
13	4.87	15	11.5	4.8	15.5	11
10	3.4	13.5	11.5	3.31	14	12.75
10						
10	3.55	14.5	11.5	3.4	14	12.5
9	3.3	14	12	4.06	15	12.25
Mean	3.96	14.3	11.6	3.83	14.5	11.9
S	0.76	0.57	0.22	0.62	0.71	0.84
t	0.305	0.492	0.771			
P	0.8 0.7	0.7 0.6	0.5 0.4			

S = Standard deviation.
P = probability that the populations do not differ.

criteria of Spinage (1975). This showed that the difference in weight of the maxillary tooth row between male and female was not significant at the 95 per cent level of probability, but length of the tooth row was (see Table 1). When the first molar was examined, mean weight, crown height, crown width and crown length, were not significantly different at the 95 per cent level: but crown length showed the greatest difference, being slightly longer in the female (Table 2). When samples of equivalent crown height were compared, it was found that weight of the first molar was greater in the male, but crown length and width were slightly smaller than in the female. But no differences were significant at the 95 per cent level (Table 3). Thus although the samples suggested that slight differences between male and female dentition may exist, the variability among teeth of either sex appeared to be greater than the difference between sexes, to the extent that the latter could be ignored in this species.

DISCUSSION

Usefulness of the single measurement of crown height

Several workers have relied upon multiple molar measurements to eliminate possible individual variation in tooth size. Robinette *et al.* (1957) proposed a 'molar tooth ratio' for North American mule deer *Odocoileus Hemionus* Rafinesque, which was derived from the sum of the occlusal widths of the seven buccal crowns, divided by the sum of the corresponding lingual crown heights. Erickson *et al.* (1970) found that this gave no better security of prediction than did visual estimation from the wear pattern; giving only 62 per cent agreement with ages derived from cementum line counts, compared with 63 per cent from visual appraisal. Multiple measurements may eliminate errors due to a few aberrantly large or small teeth, but their use entails the danger of introducing greater variability than when a single dimension is used. However, in the case of some species of North American deer, the structure of the molar teeth is such that the cross-sectional surface area increases markedly with age; and measurements of this increase may assist in the determination of age. Measurement on the relatively unworn tooth is difficult for the first maxillary molar, due to its marked trapezoid nature in cross-section, and the prominent mesostyle in some species. Crude approximations were however made of the cross-sectional areas of the maxillary and mandibular first molars at the apex of the crown, and at the neck, for several species. In the mule deer there was an increase in cross-sectional area of 23 per cent for the maxillary molar, and 21 per cent for the mandibular molar. The white-tailed deer *Odocoileus virginianus* Zimmerman showed increases of 17 per cent and 2.3 per cent respectively. It would be expected that the giraffe *Giraffa camelopardalis* L. which has cervid-like brachyselenodont teeth, would similarly show an increase; but it proved to be anomalous in this respect, with decreases of 0.8 per cent and 0.7 per cent for maxillary and mandibular teeth respectively. But the giraffe differs from the other species examined in having much larger mandibular teeth in cross-sectional area, relative to the maxillary teeth. A typical hypsoselenodont grazer, the buffalo, had an increase of 7 per cent for the maxillary molar, and a decrease of 3 per cent for the mandibular molar. The hypsoselenodont Grant's gazelle, a semi-browser, showed decreases of 2.4 per cent and 8.4 per cent. When the maxillary cross-sectional areas were combined with their mandibular counterparts, it was found that the mule deer and white-tailed deer still showed distinct increases in area with age (as expressed by the cross-sectional area at the neck of the tooth); while the remaining species probably showed no change at all. Such differences as were revealed are probably attributable to measurement errors (see Table 4 and Figure 1). This preliminary examination shows how the structure of teeth varies among species, and how a study of this structure can determine which dimensions may be useful in the estimation of age. It also points to the fact that what may be useful for North American species, may not necessarily pertain to African.

Some workers have advocated enamel height as a suitable measurement, rather than crown height measured from the interradicular apex as used here. In the Grant's gazelle there is a fairly constant difference between the two measurements ($\bar{x} = 1.175$mm, S.D. = 0.467, range 0.5–2.0); but crown height

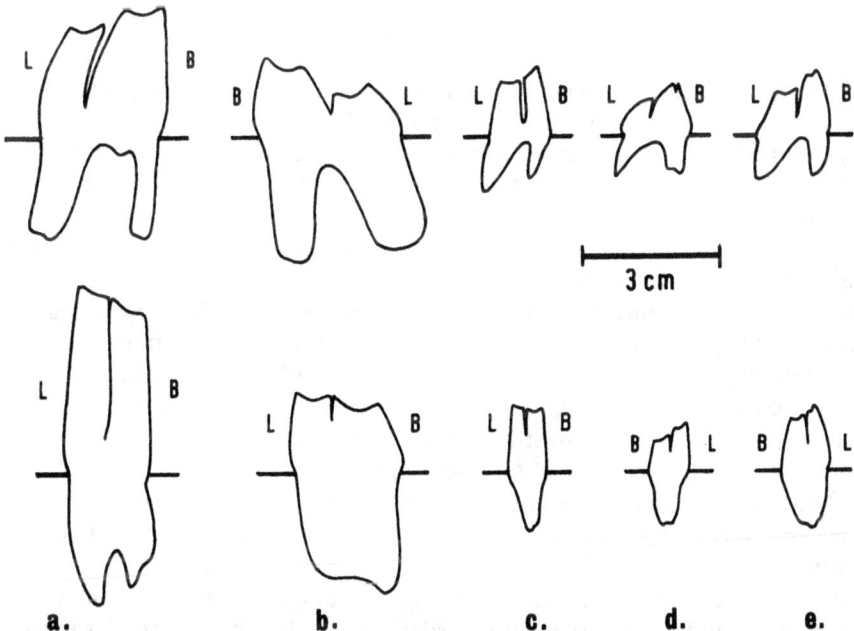

Fig. 1. Lingual-buccal sections of various maxillary (top) and mandibular (bottom) first molar teeth, with neck region indicated; showing variations in increasing width of crown with age. a = buffalo; b = giraffe; c = Grant's gazelle; d = white-tail deer; e = mule deer. L = lingual, B = buccal.

TABLE 4

*Comparison of the cross-sectional areas of first molars — crown width × crown length
(approximate measurements in mm²)*

Species	Area at crown apex		Area at neck				% Change at neck
	Maxilla	*Mandible*	*Maxilla*	*Mandible*	*Maxilla*	*Mandible*	
Mule deer	187	139	230	168	+23	+21	+22
White-tailed deer	180	132	210	135	+17	+ 2.3	+10.6
Buffalo	683	371	635	360	+ 7	− 3.0	− 5.6
Giraffe	714	595	708	591	− 0.8	− 0.7	− 0.8
Grant's gazelle	165	119	169	109	− 2.4	− 8.4	− 2.1

is preferred, as the lower enamel is often covered with cementum which requires removal before measurement. It also relates to the theory of crown height being worn to zero, and although enamel height may be worn to zero, this precedes physiological extinction. It is also preferable to take the measurement between the two cusps rather than measuring the lingual crest height of each cusp and taking the mean, as the crests are sometimes chipped rather than worn away.

Test of the model

There is a growing tendency to dress up models as if they are possessed of statistical validity; but although a model may fit the theory it may not fit reality. In this model, although the measurement of wear must reflect age, it may not necessarily predict age with consistency; either within species or among species. To test this we require a sizeable sample of known age specimens; a rarity among African mammal collections, if any exists at all. The only suitable sample of measurements from known age animals which I have found, is provided by Severinghaus (1949) for the white-tailed deer. Severinghaus measured lingual crest height above the gum, but this will not affect the model significantly; the height of the crests above the mid-point and the amount of crown concealed by the gum, probably cancelling each other out. Using measurements from 37 known-age animals the model was calculated with a Y_0 was only 1.26× the height of the tooth in the first year class, compared with 1.8× for an animal like the Grant's gazelle (Spinage, 1975), and the slope of the curve was correspondingly flatter. This accords with the brachyselenodont nature of the teeth of *Cervidae*. The end-point of 23 years followed Flower (1931) who considered this to be the potential longevity of medium and large deer, although the longest-lived recorded white-tails are two females each of 19.5 years, one of which died accidentally (Severinghaus and Cheatum, 1956). In this test, calculated values of molar height were compared with observed heights, rather than year classes being compared. The comparison gave greater than 99.5 per cent agreement (chi² = 0.363 for 8 degrees of freedom, p =

TABLE 5

*Comparison of the observed mean cusp heights[1] of known-age white-tailed deer, with cusp
height predicted from the model*

$$y = 11 \left[1 - \frac{t}{23} \right]^{1/2}$$

Observed age class[2]	Assumed age class	Observed M1 cusp height[3]	Predicted M1 cusp height
11–13 mo.	1 yr.	9.3mm	8.7mm
20–24 mo.	2	7	8.8
2.5 yr	2.5	7.2	7.4
3.5	3.5	6.8	6.7
4.5	4.5	6	6.1
5.5	5.5	5.7	5.6
6.5	6.5	4.9	5.2
7.5	7.5	5	4.7
8.5–9.5	9.5	4.5	3.9
10.5+	10.5	4.2	3.6

Chi sq. = 0.363 with 8 d.f. p = 0.005

1, 2 and 3 = data from Severinghaus 1949.

0.005): a highly significant result. It is not yet known if the model has the same
degree of association for the African species it is intended for, but the
indications are encouraging.

CONCLUSION

There is a need for a quick and simple method of determining the absolute ages
of large mammals. Investigation suggests that the height of the first molar tooth
declines in a wide range of ungulates according to a square root function. This
relationship provides a means of predicting specific age; but since, in common
with many other biological processes, the rates of wear of teeth in a population
are likely to approximate to a normal distribution, prediction should be used
for samples of a population, rather than for individuals.

REFERENCES

Chaplin, R. E. (1971) *The Study of Animal Bones from Archaeological Sites.* Seminar
    Press, London.
Erickson, J. A., Anderson, A. E., Medin, D. E. and Bowden, D. C. (1970) Estimating
    ages of mule deer — an evaluation of technique accuracy. *J. Wildl. Mgmt,* Vol. 34,
    pp.523–31.
Flower, S. S. (1931) Contributions to our knowledge of the duration of life in vertebrate
    animals. V. Mammals. *Proc. Zool. Soc. Lond,* 1931, pp. 145–234.
Grimsdell, J. J. R. (1973) Age determination of the African buffalo, *Syncerus caffer*
    Sparrman. *E. Afr. Wildl. J.,* Vol. 11, pp. 31–53.
Robinette, W. L., Jones, D. A., Rogers, G. and Gaswiler, J. S. (1957) Notes on tooth
    development and wear for Rocky Mountain mule deer. *J. Wildl. Mgmt,* Vol. 21, pp.
    134–53.

Severinghaus, C. W. (1949) Tooth development and wear as criteria of age in white-tailed deer. *J. Wildl. Mgmt,* Vol. 13, pp. 195–216.

Severinghaus, C. W. and Cheatum, E. L. (1956) Life and times of white-tailed deer. In: *The Deer of North America.* Ed. W. P. Taylor, 57–186. Wildlife Management Institute, Washington.

Spinage, C. A. (1967) Ageing the Uganda defassa Waterbuck *Kobus defassa* ugandae Neumann. *E. Afr. Wildl. J.,* Vol. 5, pp. 1–17.

Spinage, C. A. (1971) Geratodontology and horn growth of the impala *(Aepyceros Melampus). J. Zool. Lond.,* Vol. 164, pp. 209–25.

Spinage, C. A. (1973) A review of the age determination of mammals by means of teeth, with especial reference to Africa. *E. Afr. Wildl. J.,* Vol. 11, pp. 165–87.

Spinage, C. A. (1975) Age determination of the female Grant's gazelle. *E. Afr. Wildl. J.,* Vol. 13, (In press).

Steenkamp, J. D. G. (1974) Differences in the dentition of indigenous and Hereford cattle as seen under polarized light. *Rhod. J. agric. Res.,* Vol. 12, pp. 3–17.

6

PREDATOR-PREY INTERACTIONS: A CASE-STUDY IN THE MASAI-MARA GAME RESERVES, KENYA

A. R. K. Saba

Federal Department of Forestry, Ibadan, Nigeria

INTRODUCTION

The Serengeti-Mara ecosystem with an area nearly 25 million hectares contains a variety of habitats (Grzimek, 1961; Bell, 1971; Kruuk, 1972; Taiti, 1973; Talbot and Talbot, 1963). Two of the three notable attractions of this area are the Serengeti National Park in north-west Tanzania and the Masai Mara Game Reserve (south-west Kenya), (see Schaller, 1972). The third, and by far the most publicized, is the annual migration of more than one million animals, principally wildebeest, zebra and Thomson's Gazelle, over a large segment of the region (Grzimek, 1961; Bell, 1971; Watson, 1967; Sinclair, 1970, 1973; Norton-Griffiths, 1973).

The food habits of the lion were studied on the Mara side of the area for a period of 14 months — between June 1973 and July 1974 (Saba, 1974), with a view to determining whether the Mara lions showed any preference for the migratory species at the height of their migration through the area and which of the resident ungulates were being eaten in the absence of the migratory species. Most ungulate species of the Mara Game Reserve show a distinct pattern of migration, which is closely related to rainfall. The rainfall data are presented in Figure 1. The pattern of migration of the migratory species in the Serengeti has been described by Watson (1967) and Bell (1971). The Serengeti and Loita plains are visited during the wet season, whereas the northern extension of the Serengeti National Park and sometimes, the Mara Game Reserve are visited during the dry season between August and October each year.

The proportion of the Serengeti migratory ungulates that cross into Mara is not known. For the wildebeest, Watson (1967) gave an estimate of 10–15 per cent up to 1964. There has been an increase in the Serengeti wildebeest population and the latest estimate given by Sinclair (1973) was 840 000. If Watson's ratio is maintained for the animals that move into the Mara Reserve, it means that about 80 to 100 000 wildebeest from the Serengeti may be using the Mara area during late August to early October.

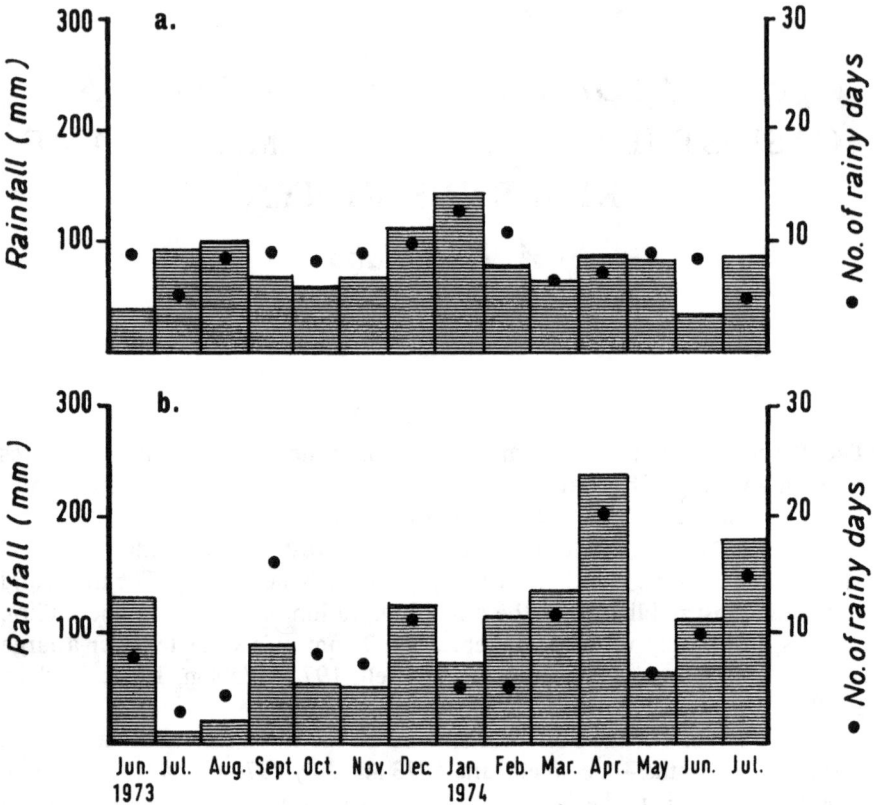

Fig. 1. Rainfall in the Mara Game Reserve.

Population figures of the other migratory species were given by Bell (1971) as 220 000 zebra and 150 000 Thomson's gazelle. If the threefold population increase found in the wildebeest has taken place among these two species, the expected population will be about 700 000 for zebra and 500 000 for Thomson's gazelle. Bradley (personal communication) has confirmed such a rise for Thomson's gazelle and gave an estimated population of 606 000 for this species in the Serengeti.

OBSERVATIONS

The study area was covered twice daily from a land cruiser.

Vulture roosts and tourist vehicles invariably indicated the location of lions and when lions were found eating, the following information on the prey species was recorded, where possible: their age and sex and the physical condition of prey, assessed from an examination of the bone marrow (Sinclair and Duncan, 1973). In the course of this work 220 herbivores were recorded eaten by lions.

TABLE 1

Composition of the diet of Mara lions

Species	Number eaten	Kill in percentages	Approximate weight (kg)
Buffalo	76	34.54	31 920
Wildebeest	75	34.09	8150
Zebra	36	16.36	5904
Giraffe	8	3.64	5728
Wart-hog	8	3.64	320
Hartebeest	7	3.18	665
Topi	5	2.27	410
Thomson's gazelle	3	1.36	36?
Eland	1	0.45	225
Water-buck	1	0.45	131
Total	220	99.98	53 489

Table 1 shows that buffalo, wildebeest and zebra account for 84.99 per cent of the total number of animals eaten by the lions.

The 220 animals represent 53 489 kg in terms of animal biomass, using the average weights given by Sachs (1967, 1968). The very large margin in favour of buffalos (in terms of weight [about 60 per cent]) is important since it suggests that lions prefer the meat of the relatively sedentary buffalo against that of the largely migratory wildebeest.

An analysis of the observed kills of the lions at different times during the study is given in Table 2, in order to assess the relative importance of the various species in the diet of the lions at different times of the year.

It seems that whenever the three most important species in the diet occurred together, more wildebeest individuals were eaten in comparison with zebra and buffalo.

The data presented in Table 3 was factorized by the 'Kruuk and Turner function' in order to obtain an estimate of the number of the different prey species in the diet during a year. The method for deriving the function (f) is given by the formula:

$$\frac{1800 \times n}{W^2} = f \text{ (Kruuk and Turner, 1967)}$$

where n is the number of lions in the population
1800 = The average weight (in kg) of meat consumed by one lion per annum.

TABLE 2

Monthly variation in numbers of animals eaten by the lions of Masai Mara Game Reserve, June 1973 to July 1974

Month	Total	Wildebeest	Zebra	Buffalo	Hartebeest	Giraffe	Topi	Thomson's gazelle	Wart-hog	Water-buck	Eland
June 1973	9	8	1	—	–	–	–	–	–		
July 1973	11	5	4	1	1	–	–	–	–		
August 1973	11	5	2	2	1	–	1	–	–		
September 1973	21	11	5	2	–	1	1	1	–		
October 1973	20	7	3	7	1	–	1	1	–		
November 1973	17	5	—	11	–	–	–	–	1		
December 1973	6	—	1	3	–	1	–	1			
January 1974	18	2	3	9	1	–	–	–	1	1	1
February 1974	46	12	9	15	–	3	1	–	6	–	–
March 1974	31	13	7	10	1	–	–	–			
April 1974	9	4	1	4	–	–	–	–			
May 1974	6	1	—	5	–	–	–	–			
June 1974	8	1	—	4	1	1	1	–			
July 1974	7	1	—	3	1	2	–	–			
Total	220	75	36	76	7	8	5	3	8	1	1

W^2 =   total weight of the animals observed eaten by lions, in kg.

For the Mara study, this means

$$f = \frac{1800 \times 129}{53489} = 4.3411$$

using the original lion population of 129.

TABLE 3

Yearly consumption of prey by Mara lions
(Kruuk and Turner, 1967)

Prey species	Number observed eaten	Calculated number consumed yearly	Estimated numbers (+1/5)
Buffalo	76	330	396 = 400
Wildebeest	75	326	391 = 400
Zebra	36	156	187 = 200
Giraffe	8	35	42 = 50
Wart-hog	8	35	42 = 50
Hartebeest	7	30	36 = 40
Topi	5	22	26 = 30
Thomson's gazelle	3	13	16 = 20
Eland	1	4	5 = 10
Water-buck	1	4	5 = 10
	220	955	1210

After Kruuk and Turner, 1967.
$f = 4.3411$: see text.

Killing rate per lion $\frac{1210}{129} = 9.4$ animals per year

The table shows the 'calculated number' of each species eaten per year by Mara lions. These figures were rounded off to 'estimated numbers' by adding 1/5 to account for wastage as suggested by the authors. This gave an estimated total of 1210 animals consumed by Mara lions during the year.

The calculated rate of 9.4 prey animals per lion per annum is then compared with the suggested range of 10–36.5 given by other authors (Table 4). The Mara figure is not significantly different from 10–12 prey per lion per annum given by Stevenson-Hamilton and 16 prey per lion per annum by Pienaar (1969).

The seemingly excessive number of prey per lion found in other areas may be related to the overall biomass of the different prey species in the areas. Whereas the heavier biomass of the prey in Mara was contributed by buffalo, which was 34 per cent of the prey consumed, there were no buffalos in the diet of Nairobi Park lions and buffalo represent only 5.51 per cent of the lion kills in Serengeti.

TABLE 4

Estimates of lion kill per annum based on different killing rates

Author	Number of kills per lion per year	Applied to Mara study area
Guggisberg (1961)	20	2580
Stevenson-Hamilton (1947)	26	3354
Talbot and Talbot (1963)	35	4515
Wright (1960)	36.5	4709
Rudnai (1970)	30	3870
Schaller (1972)	32	4128
Wells, in Guggisberg (1961)	15	1935
Kruuk and Turner (1967)	11	1419
Stevenson-Hamilton Pienaar (1969)	10–12	1290–1548
Pienaar (1969)	15	1935
This study	10	1290

It is possible therefore that the calculated killing rate in the different areas reflects the preponderance of a particular prey species in the diet, for instance Thomson's gazelle in the Serengeti and kongoni in Nairobi National Park (see Table 4). Talbot and Talbot's figure of 35 prey per lion per annum comprised largely wildebeest.

DISCUSSION

The predation records showed that wildebeest and buffalo individuals occur in about equal proportions (both nearly 70 per cent in the diet) and zebra constitute the third most important prey species in the lions' diet. The feeding pattern of Mara lions when most of the migratory species had left the study area also indicated a preference for wildebeest. As only very few wildebeest were present in the area at this time, it is suggested that some antipredator mechanism was probably operating through the very large herds in which they originally occurred at the peak of the migration. Buffaloes herding, as an antipredator device, was observed by Sinclair (1973).

Based on the number and weight of each species in the diet, the Mara lions are found to be the second lion population in Africa whose diet has a higher proportion of buffalo (see Table 5). Lions in Lake Manyara National Park have a diet that was made up of 60 per cent buffalo (Makacha and Schaller, 1969). In Kafue National Parks, buffalo was reported to be over 30 per cent in the lions' diet (Mitchell *et al.*, 1965).

This study supports Bourliere's (1963) opinion that lions feed mainly on medium-sized ungulates such as wildebeest, zebra, topi, water-buck, or submedium species like the Uganda kob, impala, and wart-hog. The elephant,

Comparison of lion food in six African reserves

Species	Manyara Park 1967–69 Schaller 1972	N'bi Park 1954–66 Foster and Kearney, 1967	N'bi Park Rudnai, 1970	Kafue Park 1960–63 Mitchell et al., 1965	Kruger Park 1954–66 Pienaar, 1969	Serengeti Park Schaller, 1972	Masai Mara this study
Number of kills	(100[d])	257	116	410	12 313	1180	220
Wildebeest	2.02	48.32	25.02	8.12	23.62	34.72	34.12
Zebra	16.0	20.7	14.6	7.3	15.8	21.6	16.4
Impala	11.0	2.7	0.9	2.0	19.7	0.3	0.5
Water-buck	1.0	trace	—	5.9	10.5	0.08	0.5
Eland	—	1.3	1.7	2.9	0.5	1.1	3.2
Hartebeest	—	10.5	34.5	16.3	—	0.5	3.6
Wart-hog	—	9.7	6.9	9.5	—	1.8	3.6
Giraffe	2.0	3.9	1.7	—	1.9	0.8	3.6
Buffalo	62.0	—	—	30.5	3.9	5.5	34.5
Bushbuck	—	—	—	0.2	0.3	0.08	—
Topi	—	—	—	—	—	2.9	2.3
Bushpig	—	—	—	2.0	trace	—	—
Duiker	—	—	—	0.2	0.1	—	—
Hippopotamus	—	—	—	1.4	trace	—	—
Kudu	—	—	—	1.0	10.9	—	—
Lechwe	—	—	—	0.5	—	—	—
Puku	—	—	—	1.0	—	—	—
Reedbuck	—	—	—	2.0	0.3	0.6	—
Roan	—	—	—	5.6	0.3	—	—
Sable	—	—	—	5.1	1.5	—	—
Tsessebe	—	—	—	—	0.4	—	—
Tommy	—	—	0.9	—	—	28.1	1.4
Granti	—	—	5.2	—	—	0.9	—
(Small antelope[a])	—	—	—	—	trace	—	—
Baboon	6.0	—	2.6	—	trace	—	—
(Carnivores[b])	—	2.3	—	—	0.4	0.4	—
Ostrich	—	—	—	—	0.1	0.3	—
Porcupine	—	—	—	0.5	0.1	—	—
(Others[c])	—	—	6.0	—	0.3	—	—
No prey species	7	9	11+	19	38	22	10
Killing rates	—	—	30	—	15	32	9.4

(a) Steen-buck, grysbuck, klipspringer.
(b) Lion, leopard, hyena, cheetah, jackal, civet, ratel, caracal.
(c) Nyala, white rhino, aardvark, pangolin, crocodile, tortoise.
(d) Three persons who were eaten by lions just outside of the park are not included.
(e) Guinea fowl, sand grouse, saddle bill stork, hare, pangolin.

the rhinoceros and the hippopotamus are known to be ignored (Schaller, 1972) or taken only under exceptional circumstances (Bourliere, 1963; Schaller, 1972).

Other important factors that are supposed to influence prey capture include vulnerability of the prey (Bertram, 1973) and aspects of the behaviour of both prey and predator (Walther, 1969; Ewer, 1973). Bertram (1973) was also of the opinion that the 'catchability' of a prey may not bear a direct relationship to its abundance. He is supported in this view by Ewer (1973) who stated that a less numerous species may be more vulnerable if they lack the protective behaviour of other species.

The efficiency of the antipredator behaviour of a prey species, as suggested by Schaller (1972) may be reflected in the hunting success of the predator. He gave figures which showed that the vigilance, agility and speed of some prey species are related to their numbers in the lions' diet.

The utilization of buffalo meat by Mara lions, particularly in the absence of migratory ungulates, indicate a notable difference between the food habits of these lions and those of the Serengeti. Schaller (1972) observed that, between migrations, Serengeti lions prey virtually on Thomson's gazelle.

It is expected that this important difference in the diet of Mara and Serengeti lions might be related to the reported killing rates of the lions. A killing rate of 9.4 ungulates per lion per annum in the Mara as opposed to a rate of 32 for Serengeti lions reflects merely the weight differences of the most abundant prey in an adult buffalo is about 35 times as heavy as an adult Thomson's gazelle.

It can be observed that where buffalo numbers in the lions' diet are minimal the predation rates are reported to vary from 15 (Pienaar, 1969) to 32 (Schaller, 1972).

The continued utilization of buffalo by Mara lions can be ensured only with the cessation of woodland destruction and other habitat despoilation. The study has shown that wildebeest and zebra represent a buffer food source for the lions.

To sustain a tourist interest in game viewing areas, Watson (1967), Kruuk (1972), and Schaller (1972) have advocated the maintenance of a healthy predator population, because, in the final analysis, population management is best handled by the predators.

REFERENCES

Bell, R. H. V. (1971) A grazing ecosystem in the Serengeti. *Scientific American*, Vol. 224, pp. 86–93.

Bertram, B. C. R. (1973) Lion population regulation. *East African Wildlife Journal*, Vol. 11, Nos. 3 and 4, pp. 215–25.

Bourliere, F. (1963) Specific feeding habits of African carnivores. *African Wildlife*, Vol. 17, No. I, pp. 21–7.

Bradley, R. M. (1970) Ecology of Thomson's gazelle. *(Serengeti Research Institute, Annual Report).* Pp. 33–6.

Croze, H. (1974) The Seronera bull problem II; the trees. *East African Wildlife Journal,* Vol. 12, No. 1, pp. 29–47.

Ewer, R. F. (1973) *The Carnivores.* (Weidenfeld and Nicolson, London).

Foster, J. and Kearney, D. (1967) Nairobi National Park Game census. *East African Wildlife Journal,* Vol. 6, pp. 152–4.

Grzimek, B. (1961) *Serengeti shall not die.*

Guggisberg, C. A. W. (1961) *Simba.* (Howard Timmius, Cape Town).

Kruuk, H. (1972) *The spotted hyaena. A study of predation and social behaviour.* (University of Chicago Press).

Kruuk, H. and Turner, H. (1967) Comparative notes on predation by lion, leopard, cheetah and wild dog in the Serengeti area, East Africa. *Mammalia,* Vol. 31, No. 1, pp. 1–27.

Makacha, S. and Schaller, G. (1969) Observations on lion in the Lake Manyara National Park, Tanzania. *East African Wildlife Journal,* Vol. 7, pp. 99–103.

Mitchell, B. L., Shenton, J. B. and Uys, J. C. (1965) Predation on large mammals in the Kafue National Park, Zambia. *Zoological Africa,* Vol. 2, pp. 299–318.

Murdoch, W. W. (1969) Switching in general predators; experiments on predator specificity and stability of prey population. *Ecological Monographs,* Vol. 39, pp. 335–54.

Norton-Griffiths, M. (1973) Counting the Serengeti wildebeest using a two-stage-sampling. *East African Wildlife Journal,* Vol. 11, pp. 135–49.

Pienaar, U. de V. (1969) Predator-prey relations amongst the larger mammals of the Kruger National Park. *Koedoe,* Vol. 12, pp. 108–76.

Rudnair, J. (1970) Social behaviour and feeding habits of lion (*Panthera leo massaica* Neumann) in Nairobi National Park. (M. Sc. Thesis, University of Nairobi).

Saba, A. R. K. (1974) The ecology of lion *(Panthera leo massaicus)* in the Masai Mara Game Reserve, Kenya. (M.Sc. Thesis, University of Nairobi).

Sachs, R. (1967) Live weights and body measurements of Serengeti game animals. *East African Wildlife Journal,* Vol. 5, pp. 24–36.

Sachs, C. (1968) A survey of parasitic infestation of wild herbivores in the Serengeti region in Northern Tanzania and the Lake Rukwa region in Southern Tanzania.

Schaller, G. B. (1972) *The Serengeti Lion.* (University of Chicago Press).

Sinclair, A. R. E. (1970) Studies of the ecology of the East African Buffalo. (D. Phil, Thesis. Oxford University).

Sinclair, A. R. E. (1973) Population increases of buffalo and wildebeests in the Serengeti. *East African Wildlife Journal,* Vol. 11, No. 1, pp. 93–107.

Sinclair, A. R. E. and Duncan, P. (1972) Indices of condition in Tropical ruminants. *East African Wildlife Journal,* Vol. 10, pp. 143–9.

Stevenson Hamilton, J. (1954) *Wildlife in South Africa.* (Cassell, London).

Taiti, S. W. (1973) A vegetation survey of the Masai Mara Game Reserve, Narok District. (M.Sc. Thesis, University of Nairobi).

Talbot, L. and Talbot, M. (1963) The wildebeest in Western Masailand East Africa. *Wildlife Monographs,* No. 12.

Walther, F. (1969) Flight behaviour and avoidance of predators in Thomson's gazelle (*Gazella Thomsonii* Gunther). *Behaviour,* Vol. 34, No. 3, pp. 184–221.

Watson, R. M. (1967) The population ecology of the wildebeest (*Connochaetes taurinus albojubatus* Thomas) in the Serengeti. (Ph.D. Thesis, Cambridge University).

Wright, B. (1960) Predation on big game in East Africa. *East African Wildlife Journal,* Vol. 24, No. 1, pp. 1–15.

7

A SAMPLE DRIVE COUNT OF LARGER MAMMALS BY MEANS OF SYSTEMATIC BELT TRANSECTS

L. P. van Lavieren

Ecole de Faune, Garoua, Cameroun Republic

INTRODUCTION

Little is known about the size of wildlife populations in West African national parks and other areas where wildlife management is practised. Existing figures are mere guesses with no indication on the degree of precision and/or accuracy. Continuous monitoring of the size of wildlife populations is a basic requirement for proper management, and population counts should be made at least once a year.

Perhaps most urgently needed, are inventories of the numerous hunting areas surrounding the majority of West African national parks. At present, hunting quotas are often fixed by trial and error. Since the revenue from hunting fees in most West African countries is an important income, both deterioration of the hunting zone by over-hunting and fixed quotas well below the potential of the area are economically undesirable.

At the moment, most wildlife population censuses in East, East-Central and South Africa are carried out from the air. In West Africa, however, the use of light aircraft or helicopters in wildlife censusing is remote. Budgets allocated to wildlife departments, if they exist, are generally too small to maintain a light aircraft on a full-time basis. In the few cases where light aircraft is available, there is often a lack of competent personnel, additional equipment and maintenance facilities.

Aerial censuses have been proved valuable on open grasslands and flood-plains where the visibility of larger mammal species is good (Norton-Griffiths, 1973, Bell *et al.*, 1973). On flood-plains, particularly, aerial censusing is the only way to obtain reasonably accurate population estimates of larger mammal populations and their distribution in a relatively short time. The same may be said of the vast wildlife areas in the open steppe vegetation of the West African Sahelien zone.

However, a much lower visibility is encountered in the Guinea and Sudan savannah woodland zone, in which West Africa's most important wildlife

50

concentrations are found. The use of light aircraft in these areas, where the physiognomy is comparable with the 'miombo' *(Brachystegia)* woodlands of Zambia and Tanzania, is limited and restricted to the brief period during the mid dry season, when most trees have lost their leaves, most grass cover has been burned and the distribution range of most animals is considerably reduced, as they become more concentrated around permanent water. Beyond this period, an aerial census would require a large number of transects of narrow strip width, but still, the accuracy of the estimates would most likely remain poor. For these reasons, there is need for ground censusing methods, which give population estimates of acceptable accuracy, and which can be carried out by available personnel at low cost.

The objective of this paper is not to review and evaluate the existing ground census techniques. It gives a possible solution to the problem of wildlife inventory in savannah woodland. The method described below is based on the simple principle of drive counts. Drive counts are still used in temperate areas for censusing deer and roedeer. Usually these involve a large number of beaters and observers. A straight front of beaters moves across the area, eventually forcing the animals out of the area, or making them back through the line of beaters. Animals driven out of the area are counted by observers stationed around the periphery of the area. Animals passing back through the line of beaters are counted by the beaters themselves. A description of this method is given by Adams (1938).

However, drive counts of this sort can only be carried out on areas of limited size, preferably bounded by rivers, roads, fences or fire breaks. Most West African parks have a large area so that the application of this method would require a large number of beaters and observers, which is not practicable in most cases.

Using a sample drive count reduces the manpower since observers do not need to be stationed around the periphery of the area. This method was used in early 1974 by students of the College of Wildlife Management for francophone African countries, Garoua (Cameroun Republic), in the Bouba Ndjida National Park, Cameroun Republic. Their results are compared with density estimates in that area obtained by another method.

The Boubandjida National Park is situated in the north-east of the Cameroun Republic, between 9°.00′ and 8°.21′ N.; and 14°.25′ and 14°.55′ E. The 2140 sq. km of undulating country lies at an altitude varying between 280 m (riverbanks) and 360 m, with a number of rocky outcrops ranging from 500–900 m above sea level. Bosch (1975) has defined eight main vegetation types. For census purposes, he suggests dividing the park into two levels — of high and low animal densities. The central region of the park, in which the census was carried out, lies largely in the high density level and is covered with *Terminalia laxiflora* savannah woodland, which Bosch considers the best wildlife habitat in the park. All the sides of the 111 sq. km area are bounded by roads.

METHODS

Fifteen transects were spaced systematically, approximately 1 km apart along the southern border road of the area which runs east–west. Each transect was counted by a team of 3 students: a left and right observer/beater 75 m on either side of the central observer/beater. In this way, 15 belts of 150 m width were counted. All students followed the compass bearing due north and started simultaneously. The left and right observer/beater noted all animals crossing the line of march to their left and right respectively. The central observer/beater noted all animals which ran back through the line between himself and the left and right observers/beaters. He also checked, at five minute intervals, the belt width to his left and right by pacing. The three students remained visible to each other throughout the census and moved slowly ahead in one front. Transect lengths varied from 5 to 7 km and a belt width of 2×75 m was chosen, based on results of visibility measurements taken in the same area. The mean visibility along 14 transects was measured prior to the census, using students in dark-green uniform moving away perpendicularly, to either side of the transect line (Lamprey, 1964; Hirst, 1969; Robinette et al., 1974).

The calculated 'mean visibility' of a standing and kneeling student are given in Table 1. As can be seen from this table, 75 m between the students is unlikely to allow many animals to escape unseen.

However, animals may go unnoticed ahead of the students. This source of

TABLE 1

Mean visibilities (left and right) along 14 transect lines of standing and kneeling student in dark-green uniform in an area of the high density stratum, Bouba Ndjida National Park, Cameroun Republic

Transect	Standing			Kneeling		
	n	Mean visibility (m)	S.D.	n	Mean visibility (m)	
1	26	120.8	49.04	26	107.3	36.72
2	55	165.0	90.60	56	108.1	63.44
3	15	134.0	38.28	16	113.4	39.06
4	46	144.0	39.51	46	124.0	25.09
5	40	167.5	59.43	40	139.4	58.53
6	52	152.9	82.87	51	114.1	54.13
7	60	131.2	50.60	60	90.8	40.34
8	40	158.8	48.22	40	99.4	24.99
9	29	176.2	73.90	28	95.4	56.06
10	65	149.0	78.28	64	108.2	47.31
11	24	154.8	86.29	24	74.4	65.43
12	24	166.7	88.05	24	110.4	54.63
13	34	152.4	91.36	34	115.4	71.78
14	14	108.9	37.48	12	93.8	32.20

n: number of points measured, S.D.: standard deviation.
N.B. visibility of standing student taken for bubal hartebeest and roan, visibility of kneeling student taken for oribi, etc., (see Table 2).

error is very important in areas of dense vegetation growth or tall grass cover where the sample drive count should not be used. However, where the grasscover has been burned, visibility is greatly augmented and it was found that the fleeing distance from walking human beings was often considerably less than the measured maximum visibility of the species mentioned below.

Another source of possible error are animals leaving one belt and passing into the next, resulting in a double count. For this reason, transects are spaced at least 1 km apart and all drivers start simultaneously, moving forward at approximately the same speed. Moreover, the time of an animal or herd sighting is recorded, which can be helpful if doubt arises as to whether animals have been counted twice. To our knowledge this occurred only once with a small group of giraffe.

Only the following animal species were considered in determining densities: oribi *Ourebia ourebi* (Zimmermann), Bohor reedbuck *Redunca redunca* (Pallas), common duiker *Sylvicapra grimmia* (L.), Bubal hartebeest *Alcelaphus buselaphus* (Pallas), and roan antelope *Hippotragus equinus* (*Desmarest*).

RESULTS

The results are given in Table 2. They have been compared with the results of 14 mean visibility transects (Lamprey, 1964). Table 3 shows the sample size used in both methods. The differences between the calculated means of both methods have been tested with the Student or t-test. Results are given in Table 4. Using 5 per cent confidence limits the densities were not found to differ significantly.

TABLE 2

Results of 15 belt transects (sample drive count) compared with results of 14 'mean visibility' transects (Lamprey, 1964) in the same parcel of the high density stratum, Bouba Ndjida National Park, Cameroun Republic

Species	Sample drive count belt transects (Feb. 1974)			Mean visibility method (Mar. 1974)		
	n	*Mean density per sq. km*	*S.E. of transect means*	*n*	*Mean density per sq. km*	*S.E. of transect means*
Oribi	123	7.29	0.95	111	5.84	0.96
Reedbuck	87	5.68	1.04	61	3.95	1.13
Common duiker	30	1.94	0.40	17	1.00	0.27
Bubal Hartebeest	88	5.73	1.75	115	4.68	1.19
Roan	65	4.20	1.51	78	3.54	1.42

n: number of animals seen in transects.
S.E.: Standard Error.

TABLE 3

Sample sizes of sample drive count and mean visibility method in an area of high density stratum, surface area of study area is 110.6 sq. km

Method	Number of transects	Sampled area sq. km	Percentage of the area
Sample drive count	15	14.895	13.5
Mean visibility method Bubal hartebeest, roan and oribi, etc.	14	24.510 17.912	22.2 16.2

TABLE 4

Testing differences of the calculated means of both methods, see Table 2, using the student-test (5 per cent confidence limits, 27 degrees of freedom)

Oribi	$t = 1.07 < 2.052$	not significantly different		
Reedbuck	$t = 0.93 < 2.052$	,,	,,	,,
Common duiker	$t = 1.96 < 2.052$	,,	,,	,,
Bubal hartebeest	$t = 0.50 < 2.052$	,,	,,	,,
Roan	$t = 0.36 < 2.052$	,,	,,	,,

One should not be too optimistic with the results obtained from either aerial or ground census techniques in African savannah woodland. Papers on this subject often show remarkably precise figures, but the degree of accuracy often remains undiscussed.

The degree of accuracy reflects the ability of the observer's eye to spot animals. Since the eye does not register the ones overlooked, this degree remains unknown. Accuracy is only really measurable, if the sample count is carried out on populations of known sizes. As a result, accuracy is often sacrificed for precision. If the census is carried out to determine trend in population size, consecutive counts of maximum precision with a constant degree of accuracy are desired (Caughley, 1972). This means that consecutive counts should be carried out using the same method, under the same conditions (time of the year, time of the day, weather etc.) and preferably, by the same observers. For this reason the method described here could be used in open savannah woodland of homogeneous composition. Results obtained with this simple method did not differ from the results of another method which has been applied successfully elsewhere.

Before carrying out a wildlife census, special attention should be given to stratification of the area. This should take into account the following:

(1) The distribution of animals during the time of the year when the census was carried out.

(2) The density of animals in the different areas of the park in relation to the quality of habitat, water availability etc.

(3) The vegetation cover in relation to visibility.

These data are often not known for West African wildlife areas, and therefore, as in the above case where only two strata have been suggested, stratification must remain less detailed.

Suggesting a 10 per cent sample in the high density stratum and a 5 per cent sample in the low density stratum, a park, the size of Bouba Ndjida, could be censused by 40 men in less than 2 weeks using the method presented in this paper.

ACKNOWLEDGEMENTS

I would like to thank students of year 1973/74 of the Garoua Wildlife Management College for their assistance in this census. I also acknowledge the useful suggestions made by M. Bosch who is currently carrying out a research programme in the area.

REFERENCES

ADAMS, H. E. (1938) Deer census and kill records of the Lake States. *Trans. N. Amer. Wildl. Conf.* 3, 287–95.

BELL, R. H. V., GRIMSDELL, J. J. R., VAN LAVIEREN, L. P. and SAYER, J. A. (1973) Census of the Kafue lechwe by a modified method of aerial stratified sampling. *East African Wildlife Journal*, Vol. 11, pp. 55–74.

BOSCH, M. (1975) Progress report of research in Bouba Ndjida National Park. Unpublished.

CAUGHLEY, G. (1972) Aerial survey techniques appropriate to estimating cropping quota's. SF/FAO 26 proj. Work. Doc No. 2.

HIRST, S. M. (1969) Road-strip census techniques for Wild Ungulates in African woodland. *Journal of Wildlife Management*, Vol. 33, No. 1, pp. 40–8.

LAMPREY, H. F. (1964) Estimation of the large mammal densities, biomass and energy exchange in the Tarangire Game Reserve and the Masai steppe in Tanganyika. *East African Wildlife Journal*, Vol. 2, 1–46.

NORTON-GRIFFITH, M. (1973) Counting the Serengeti migratory wildebeest by using two-stage sampling. *East African Wildlife Journal*, Vol. 11, No. 2, pp. 135–50.

ROBINETTE, W. L., LOVELESS, C. M. and JONES, D. A. (1974) Field tests of strip census methods. *Journal of Wildlife Management*, Vol. 38, No. 1, pp. 81–96.

8

VETERINARY ASPECTS OF ECOLOGICAL MONITORING

M. H. Woodford

UNDP/FAO Kenya Wildlife Management Project KEN/71/526

INTRODUCTION

Current interest in ecological monitoring is revealing many instances where the data which are gathered can be interpreted and used by diverse disciplines.

The concept of the natural nidality of transmissible disease was proposed by Pavlovsky in 1939. This Russian academician, however, was concerned with those diseases which occur naturally in wildlife populations and are transmitted to man, by way of arthropod vectors, when he enters the territory shared by the reservoir host species and the vector.

It has, of course, long been recognized, both in Russia and the West, that many diseases of man and his domestic animals have reservoirs in wildlife and that arthropod vectors are often involved in transmission.

There is usually a firm association between biotic factors such as the disease agent, the reservoir host, the vector and the vegetation, on the one hand, and the abiotic such as geographical landscape, rainfall and soil type, on the other. The landscape may be virgin, settled or greatly changed by man and each of these circumstances may produce a different set of veterinary/medical problems.

Very often the epidemiology of a specific disease undergoes a marked change in the habitat between two landscapes or at the interface between adjacent forms of land-use.

Vector-borne infections of man and animals which are maintained in wild reservoir hosts are particularly likely to have well-defined geographical limits. These are determined by environmental factors, especially climate and soil type, which in turn influence vegetation and thus the spatial distribution of the wildlife.

Movements of wild animals in annual migrations and domestic animals in treks to grazing areas, water-holes and to market also provide opportunities for the spread of disease agents and their arthropod vectors (Hoogstraal, 1972).

For the purpose of this paper it would be well to define some of the terms used.

A *reservoir host* is an organism, vertebrate or invertebrate, which is infected with an infectious agent whereas a *donor* is an infected animal currently suffering from a viraemia, bacteraemia or parasitaemia which is thus able to serve as a source of infection for a vector.

A *vector* is an invertebrate organism (usually an arthropod or insect) which acquires a disease agent while feeding on an infected vertebrate (human or animal) and later transmits it to another host. Transmission may be mechanical as when tabanid flies transmit *Trypanosoma evansi* from one camel to another, or cyclical as when the tick, *Rhipicephalus appendiculatus* transmits the protozoon, *Theileria lawrencei* from a buffalo (*Syncerus caffer*) to a domestic cow.

Zoonoses are those infections which are naturally transmitted between vertebrate animals and man.

In this paper biologically heterogeneous disease conditions will be considered with special emphasis upon those whose natural foci can be detected or deduced by an examination of ecological monitoring data. Some of the diseases are zoonoses, some are confined to domestic stock and some to wildlife. All are of social and economic importance and must be considered when land-use decisions have to be made.

Ecological monitoring techniques can provide data which can be used:

(1) to define the geographical limits of important diseases of human beings and their livestock,

(2) to determine why such spatial limits exist and what controls them, and

(3) to predict the likely veterinary/medical consequences of changes and trends in current land use systems.

The various categories of ecosystem data which will be made available by an Ecological Monitoring Unit have been described by Gwynne and Croze (1975). These are divided into environmental, faunal and politico-economic categories but only the first two will be considered here.

ENVIRONMENTAL FACTORS

Permanent Attributes

Topography

Elevation, and its effect on rainfall and maximum and minimum temperature, places a marked constraint on the distribution of potential wildlife disease reservoirs and arthropod vectors. Where the severity of the climate is seasonal there may be a vertical movement of susceptible and reservoir stock across the contour lines.

Soils

The chemical content of soil substrates may be associated with various diseases e.g. fluorine, selenium (Rosenfeld and Beath, 1964) and arsenical poisoning, and with iodine and copper deficiencies.

Soil structure may limit the distribution of hookworm and other parasitic nematodes (Augustine and Smillie, 1926).

Drainage

Drainage lines and drainage barriers affect vegetation and thus the micro-climate. They provide the habitat of the molluscan intermediate hosts of flukes, schistosomes and some lung worms and the breeding places of insect vectors such as mosquitoes and simuliids. Rising and falling water-tables influence salination which kills some vegetal types and favours others. Thus bush may die out when the water-table rises and the habitat of certain tsetse species (vectors of human and animal trypanosomiasis) can be so altered as to eliminate these insects.

Swampy areas which arise from impeded drainage, streams and rivers all provide habitats for various species of aquatic rodent, some of which are the reservoir hosts of *Leptospira* and *Tularaemia*.

Man-made environmental alterations comprising dams, irrigation systems and canals can all have profound influences on the ecology of pathogenic agents, their reservoirs and vectors.

Water-holes

Natural springs may be mineralized, boiling hot in volcanic areas or icy cold when fed by glaciers. Springs often feed swamps and the environs of such areas can become concentration points for anthrax spores and Johnes disease bacilli.

Man-made dams and boreholes, natural water-holes and pans dry up and as they do so often provide the ecological requirements of anthrax incubation — pH 7.5, a mean water temperature of 27°C and a high organic content (Ebedes, 1974).

Boreholes may tap underground sources which are heavily mineralized with toxic amounts of heavy metals, e.g. lead and arsenic (Chen *et al.*, 1962).

Water from artificial boreholes, pipelines and capped springs is often available only to domestic stock and may in fact deprive wildlife of a source previously used by them.

Trampling of a wide radius round water sources (natural and artificial) seriously affects water penetration by compaction of the surface soil. This, coupled with over-grazing, intensifies run off with far-reaching consequences for adjacent pans and swamps. Vegetation and thus micro-climates are profoundly influenced by over-grazing and trampling and such areas are concentration points for arthropod vectors and disease agents. Stock trails to and from water present similar epidemiological hazards.

Caves

Caves occur in a variety of geological formations and provide a day roost for colonies of bats, a night roost (usually near the entrance) for diurnal birds and a refuge for hyenas, ratels and other mammals. Bats and birds are a food source

for soft ticks and in some countries bats and other mammals are reservoirs of
rabies virus.

Static animal features

Termite mounds provide shelter for snakes and viverrids (reservoirs of rabies)
and breeding places for *Phlebotomus* spp. (sand flies), the vectors of
leishmaniasis. They also act as indicators of certain soil types and drainage
patterns. Other animal burrows (wart-hogs and ant-bears) are the foci of
infestation with fleas (plague) and soft ticks (African swine fever, relapsing
fever).

Bird roosts and nesting colonies recognizable by the profusion of white
droppings on the trees and ground are often the biotope of soft ticks of the
argas group, which may carry virus which can cause disease in man (Hoogstraal
et al., 1970). Holes in the trees may contain water in which mosquitoes, which
feed on the birds, can live. These mosquitoes are the vectors of avian malaria
and other blood parasites.

The burrows of colonial rodents like gerbils, susliks and springhares provide
the biotope of soft ticks, hard ticks, mosquitoes, sand flies and fleas which may
become the vectors of relapsing fever, brucellosis, tick spirochaetosis, plague,
leishmaniasis, tularaemia, listerellosis, toxoplasmosis, virus encephalitis and
tick ricketsiosis.

Semi-permanent attributes

Plant physiognomy (cover, vegetation type)

Landscape zones are often classified according to the vegetation type which in
turn will influence the geographical limits of animal populations and the
vectors of their pathogens. The combination of soil type, climate, vegetation
and land use may give rise to a set of circumstances which favour the
establishment of natural foci of certain infections or may delineate the limits of
plant toxicities and vector habitats.

Plant community composition

Some plant species concentrate toxic amounts of heavy metals, e.g. selenium,
while others contain poisonous alkaloids and glucosides and agents which
cause photosensitization in animals which eat them.

Zoogenic features (wallows, salt licks etc.)

Wallows fluctuate in extent seasonally and may become the graves of sick
animals, e.g. buffalo suffering from tuberculosis (Woodford, 1971). They thus
become heavily contaminated with infectious pathogens. Droplet transmission
of respiratory disease is also made possible when animals crowd together in
wallows. Salt licks can become infected with anthrax spores and can act as focal
points for the spread of this disease.

Distribution of non-migratory large mammal species

This may be important in the study of the epidemiology of certain enzootic diseases.

Human settlement (villages, roads, farms, ranches)

Local changes in landscape type due to small-scale human activities can have an effect on the spatial distribution of insect vectors, e.g. the interface between bush clearance and true tsetse habitat. Large-scale changes like the construction of irrigation ditches may have profound epidemiological effects on both animals and humans. Old buildings, cemeteries, and abandoned human habitations provide shelter for several potential wild animal disease donors (hedgehogs, rats, tortoises, snakes and lizards) which are the sources of food for *Ornithodorus* ticks and sand flies, the vectors of relapsing fever, and of leishmaniasis respectively.

Abandoned kraals and manyattas often swarm with hungry fleas (usually *Ctenocephalides felis strongylus*) and soft ticks (*Ornithodorus moubata*). The nitrogen-enriched soil of old cattle boma sites results in major and long lasting changes in the vegetation, e.g. boma sites on the Laikipia plateau in Kenya which date from the Maasai wars of Joseph Thomson's time (1885) and can still be seen.

Ephemeral or seasonal attributes

Rainfall, soil moisture, evapotranspiration

Many animals and their parasites are constrained by well defined climatic limits and the drawings of bioclimatographs based on the known ecological requirements of the animals is a useful exercise. Disease agents which must spend a stage of their life cycle exposed to climatic influences are also spatially limited by their environment.

The free-living stages of many parasitic nematode larvae are dependent for survival on quite narrow margins of maximum and minimum temperature and humidity as well as on soil conditions (Cameron, 1956; Levine, 1963, 1965). It is probable, too, that certain bacterial, fungal and viral agents have similar limits and there is circumstantial evidence that the virus of malignant catarrhal fever may survive for longer in moist than in dry conditions. Climatic factors, which are the subject of marked variations, may give rise to changes in epidemiological patterns as for example when a prolonged rainy season results in a larger than normal build up of mosquitoes and other insect vectors.

Associations such as these have predictive value in that epidemics can be forecast and control measures put into operation in advance.

Climatic extremes, too, can have a direct effect on populations and may cause mortality in the very young and very old and also in those concurrently stressed by parturition and lactation. Animals which hibernate or aestivate to avoid climatic extremes may carry pathogens over their period of sleep and be responsible for fresh outbreaks of disease when they emerge (hedgehog/foot and mouth disease).

Insolation

Sunlight acts as a sterilizing agent and many pathogens do not survive for long under intense ultra-violet irradiation. The infra-red end of the spectrum is also lethal to the eggs of some parasitic nematodes even when these are embedded in a dung pellet (Woodford, 1971, unpublished data).

Plant phenology

Plant phenology has an influence on animal movements and thus the distribution of infectious agents.

Plant productivity

Measurements of declining plant biomass can be indicative of a rodent irruption which may foretell the approach of an epizootic such as plague.

Chemical composition of plant parts may be important in that some plants concentrate certain elements (e.g. selenium and gold) and can be toxic at certain growth stages. Similarly some grass species under certain climatic conditions can produce hydrogen cyanide in toxic amounts. Energy content may assume great importance when the climatic conditions vary in the extreme and animals fail to obtain sufficient nutrients to enable them to breed or lactate satisfactorily.

Distribution of migratory large mammal species

Seasonal movements of migratory species may explain the epidemiology of some infectious diseases, e.g. the seasonal excretion of cell free MCF virus by parturient wildebeest.

Large mammal productivity

Disease factors can have a profound effect on productivity, e.g. 'yearling disease' (rinderpest) on the Serengeti (Talbot and Talbot, 1963) and some parasitic infestations of antelopes.

Large mammal population structure

Herd population structure is often unbalanced by losses of certain susceptible age groups due to extreme climatic conditions and to the selective action of disease agents on hosts of different species, age and sex. Thus rinderpest does not affect all species and age groups with the same severity and older animals which survive early contact with pathogenic agents are often protected by acquired immunity.

Fire

Fire temporarily changes the vegetal landscape and may kill off large numbers of arthropod vectors, parasitic nematode eggs and larvae. Fire produces changes in vegetation which may favour the survival of some disease agents while suppressing others. At the same time the sudden removal of vegetal cover over large areas must have a profound effect on the vulnerability of rodent reservoir hosts to predation

Surface water

Temporary flooding may sterilize large areas of termites, some tick species, small burrowing rodents, and certain plant species.

The interface between the temporarily flooded areas and the surrounding landscapes may thus become important epidemiological zones.

Flood-plains occurring seasonally in arid and semi-arid zones are important epidemiological features and often provide seasonal foci of several reservoir hosts and vectors.

TECHNIQUES OF ECOLOGICAL MONITORING

The techniques of data collection and storage, and its subsequent treatment to provide maps and trend surface analyses have been ably described by Gwynne and Croze (1975).

The types of data described above are currently gathered by three methods: ground sampling, aerial censusing (systematic reconnaissance flights) and by remote sensing.

Ground sampling

Data collected on the ground, either by human technicians or mechanically, are used to check data and as an essential adjunct to data gathered remotely by aerial survey and remote sensing. Of course these latter methods can only provide rather crude and imprecise data whereas a ground operator can be much more precise.

Permanent features such as topography, soils, drainage lines, water-holes caves, and animal and bird colonies can be checked for the presence or absence of minerals (toxic and trace), molluscan intermediate hosts, arthropod and insect vectors, salination, mammal, bird and reptile populations and their ectoparasites. The animals, themselves, and the arthropods and insects which subsist upon them can also be checked by a variety of cultural and serological methods for the presence of likely disease agents.

Semi-permanent features of vegetation cover and type should be examined on the ground in order to detect biotopes of disease reservoirs and vectors and, if present, their spatial distribution in relation to the vegetation mapped.

Zoogenic areas may be surveyed by bacterial culture for the presence of pathogens like anthrax spores. Meteorological data is of great importance in defining climatically constrained animal and disease limits. Insolation, rainfall, air temperature, humidity, soil moisture and soil temperature can all be monitored automatically at selected ground stations.

Small samples of slaughtered animals, both wild and domestic, will provide evidence of pathology, samples of blood, serum and parasites which can be further analysed at a base laboratory. Such samples, along with collections of arthropod vectors should be made at carefully selected points throughout a specific habitat so that the spatial limits of pathogens, reservoirs and vectors can be accurately defined. Seasonal fluctuations can be similarly monitored, along with estimates of body condition.

Predictions of probable disease outbreaks can also be made by monitoring specific reservoir, donor and vector, populations. This can sometimes be done indirectly, as for example when the disgorged pellets of diurnal birds of prey are searched for the fur of nocturnal rodents which have become diurnal under the influence of the plague bacillus (*Pasteurella pestis*).

Outbreaks of epidemic or epizootic disease are thought to indicate a state of ecological imbalance and the presence of endemic or enzootic disease to denote a climax or balanced state (Schwabe, 1969). Disease, however, may be a form of, or a consequence of, ecological imbalance and by monitoring 'indicator' species forecasts of changes in disease patterns can be made. Thus over-grazing by large herbivores on African grasslands may be followed by an increase in the number of wart-hogs which are the symptomless carriers of African Swine Fever virus, an agent which produces a very high mortality when it infects domestic pigs. Similarly, local extermination of leopard, a major predator on baboons, may be a factor in the great increase in the numbers of this species in the settled areas of Africa. Baboons are symptomless reservoirs of *Schistosoma mansoni*, the causal agent of bilharziasis in man and since these animals contaminate water supplies, the opportunity for epidemic spread of this disease exists (Nelson *et al.*, 1962).

An experienced landscape epidemiologist can often designate an area as 'bilharzia country' or 'leishmania country' after what may seem to a layman to be a relatively cursory examination.

However, subjective judgements of this nature can be of great value in indicating the direction of subsequent investigation in depth.

Monitoring of specific parasites of both wild and domestic animals can, by deduction, yield information on the presence within the host's range, of donors, vectors and intermediate hosts. Thus, the discovery of say, a specific larval pentastome in the bronchial lymph node of a gazelle will confirm the presence, within the herbivore's range, of the obligate final host of the parasite, in this case identifiable down to genus.

The load and mix of gastro-intestinal nematodes can be correlated with climatic conditions and host population density and structure. Cropping schemes, if selective for certain age and sex groups may upset the host/parasite balance and forewarning of this will be evident in data collected at ground sampling stations.

Analysis of blood meals taken by haematophagous arthropods and insects will reveal the presence of undetected food source species (Weitz, 1963) and the discovery of viral, bacterial and protozoal parasites in these vectors often obviates the need to collect and sample large mammal reservoir hosts.

Systematic reconnaissance flights

Relatively low-level (100 m above ground level [a.g.l.]) flights by fixed wing aircraft are used to obtain data on large mammal populations and also to measure a number of ecological parameters.

Visual observations, with transfer of data to a map or storage on a tape, may be made by a trained observer. Systematic photography in black and white, colour or false colour can be used to collect data for later analysis. Both these methods will reveal permanent and semi-permanent 'attributes' of an ecosystem and some of the ephemeral ones too.

Low-level photography of African elephant and buffalo (*Syncerus caffer*) herds (Croze, 1972; Sinclair, 1969) has been used to monitor recruitment age structure and thus can reveal the presence of diseases and conditions which affect reproductive efficiency. Domestic herds and flocks can be similarly monitored and this has been done by Western (unpublished data) who used pre- and post-drought colour photographs of Maasai herds to assess colour-morph-specific mortality rates.

Habitat information can be obtained from high (3000 m a.g.l.) level photography and over-grazing by large herbivores and by rodents in irruption can be detected on false colour photographs. This information coupled with other, often indirect indications of an impending epizootic can act as an early warning system and permit timely preventive measures.

Remote sensing

Images transmitted from a programmed satellite (e.g. an Earth Resources Technology Satellite [ERTS]) are essentially the same as aerial photographs obtained from a low-flying aircraft but due to the much greater height from which they are taken they naturally lack some detail.

They do, however, show adequately major land features — drainage lines, mountains, cliffs, faults, lakes etc. False colour is used to enhance green vegetation and an area which is drying up will show a progressive colour gradient from the bright red of healthy grass to the bright pink of a drying range. The satellite passes overhead at a height of 900 km every 18 days and subject to there being no obstructing cloud cover consecutive photographs of the same area can be obtained.

Practical application of veterinary data acquired by ecological monitoring

It is stressed that the veterinary/medical interpretation of routine ecological monitoring data is equally important for solving problems involving human, domestic animal and wildlife disease situations and the interests of a doctor of human medicine overlap with those of the veterinarian where anthropo-zoonoses and zoonoses are concerned.

From the basic monitoring data the doctor or veterinarian can prepare maps which delineate:

(1) Areas of actual disease occurrence.
(2) Areas of potential disease occurrence.
(3) At a later date, when monitoring has been in operation for a period — 'probability maps' of disease occurrence, e.g. in area X a Foot and Mouth Disease outbreak is likely in eight years in ten; in area Y a rabies outbreak is likely in one year in ten and so on.

These latter predictions have enormous practical application both for control and prophylactic measures and could alone justify the considerable effort and expense of a monitoring system.

Proper appreciation by planning authorities of what can be deduced from the ecologist's basic data will go a long way towards preventing the so-called ecological disasters, which are often the result of over-enthusiastic development and an ignorance of the consequences of upsetting finely adjusted, long-established balances.

ACKNOWLEDGEMENT

I am grateful to Drs M. D. Gwynne and H. Croze who read the manuscript and made many useful suggestions.

REFERENCES

AUGUSTINE, D. L. and SMILLIE, W. G. (1926) The relation of the type of soil of Alabama to the distribution of hookworm disease. Vol. 6, pp. 36–62.

CAMERON, T. W. M. (1958) *Parasites and Parasitism.* (John Wiley and Sons Inc., New York).

CHEN, K., WU, H. and WU, T. (1962) Epidemiological studies on blackfoot disease in Taiwan. *Mem. Coll. Med. Nat. Taiwan Univ.* Vol. 8, pp. 115–29.

CROZE, H. (1972) A modified photogrammetric method for assessing the age-structure of elephant populations and its use in Kidepo National Park. *East African Wildlife Journal*, Vol. 10, pp. 91–115.

EBEDES, H. (1974) Anthrax epizootics in Etosha National Park and an historical note on Anthrax in northern South West Africa. *S.W. Afr. Nat. Cons. and Tour. Div. Progr. Rpt.*

GWYNNE, M. D. and CROZE, H. (1975) East African Habitat Monitoring Practice: A review of methods and application.

HOOGSTRAAL, H., OLIVER, R. M. and GUIRGIS, S. S. (1970) Larva, nymph and life cycle of *Ornithodorus (Alectorobius) muesebecki*, a virus infected parasite of birds and petroleum industry employees in the Arabian Gulf. *Entomological Society of America Bulletin*, Vol. 63, No. 6.

HOOGSTRAAL, H. (1972) The influence of human activity on tick distribution, density and diseases. *Wjadomosci parazytologiczne*, Vol. 18, Nos. 4, 5 and 6.

LEVINE, N. D. (1963) Weather, Climate and the Bionomics of Ruminant Nematode Larvae. *Advances in Veterinary Science,* Vol. 8, pp. 215–62.

LEVINE, N. D. (1965) Bioclimatographs, evapotranspiration, soil moisture data and the free living stages of ruminant nematodes and other disease agents. In Rosicky, B. and Heyberger, K. (Eds) *Theoretical Questions of Natural Foci of Diseases*. (Czech. Acad. Sci., Prague).

NELSON, G. S., TEESDALE, C. and HIGHTON, R. B. (1962) The role of animals as reservoirs of bilharziasis in Africa. In Wolstenholme, G. E. W. and O'Connor, M. (Eds) *Bilharziasis*. (Ciba Foundation and Little, Brown and Company, Boston).

PAVLOVSKY, E. N. (1964) Prirodnaya ochagovost' transmissivnykh bolezney V svyazi S landshaftnoy Epidemiologiey Zooantroponozov Moskva-Leningrad. Translated in Levine, N. D. (Ed) (1966), and published as *Natural Nidality of Transmissible Diseases*. (University of Illinois Press, Urbana and London).

ROSENFELD, I. and BEATH, O. A. (1964) *Selenium. Geobotany, biochemistry, toxicity and nutrition*. (Academic Press, New York).

SCHWABE, C. W. (1969) *Veterinary Medicine and Human Health*. (The Williams and Wilkins Co., Baltimore).

SINCLAIR, A. (1969) Serial photographic methods for population, age and sex structure. *East African Agricultural and Forestry Journal*, Vol. 34, pp. 87–93.

TALBOT, L. M. and TALBOT, M. H. (1963) The Wildebeest in Western Maasailand, E. Africa. *Wildlife Monographs*, No. 12 (The Wildlife Society, Washington).

WEITZ, B. (1963) The feeding habits of Glossina. *World Health Organization Bulletin*, Vol. 28, p. 711.

WOODFORD, M. H. (1971) Tuberculosis in the African buffalo *(Syncerus caffer)* in the Queen Elizabeth National Park, Uganda. (PhD Thesis, University of Zurich).

9

DISTRIBUTION OF WILDLIFE IN RELATION TO THE WATER-HOLES IN TSAVO NATIONAL PARK (EAST), KENYA

J. S. O. Ayeni

Kainji Lake Research Institute, New Bussa, Kwara State, Nigeria

INTRODUCTION

In spite of the importance of water to animals very little research has been done to evaluate its significance to the ecology of African mammals (Young, 1970). Water which is a major requirement of all animals is supplied in part from water consumed voluntarily, from water preformed in the food and water formed within the body as a result of oxidation in the tissues.

African mammals differ in their ability to 'economize' water and these differences reflect their relative abilities to withstand aridity and to colonize habitats far from permanent water supplies. In order to maintain water balance in an arid environment, animals economize water through the production of dry faeces, concentrated urine and by reducing panting and sweating — avenues of evaporative water losses.

Various factors are known to influence animals' water intake. Different volumes of water are required by individual animals for the maintenance of normal growth, fattening, later stages of pregnancy and lactation. In the same animal, water intake varies with ambient air temperature (Finch, 1972). The physiological condition of an animal also influences the water required for the maintenance of homoiostasis (A.R.C., 1965). Water intake per unit of dry matter is higher for low levels of dry matter eaten by wildlife (Jarman, 1973). High protein foods (Taylor et al., 1969) and salty diets (Macfarlane et al., 1967) are associated with higher water intake.

In the wild, game obtain drinking-water from natural and artificial supplies. The natural sources of drinking-water are rivers, springs, lakes, and natural water-holes (pans) filled with rain-water. The artificial water supplies are boreholes, dams and reservoirs. Thus the quality and the availability of water supplies for wildlife vary according to the season and the location of the habitat.

Despite frequent management policies directed at the creation of artificial water supplies in the national parks and game reserves where natural supplies are not sufficient for wildlife, so far there have been no field studies on the use of water-holes in East Africa. Recently, the periodicity (Ayeni, 1975a) and seasonality (Ayeni, 1975b) of game visits to the artificial water-holes have been established in Tsavo National Park (East) Kenya. These studies show that wildlife collect near the artificial water-holes which they use more intensively in the dry season than in the rainy season.

During the rains the game species supposedly disperse away from the dry-season water supplies those artificial water-holes which contain water at the peak of the dry season, and drink from the rain-filled natural water-holes. The main objectives of this study were:

(1) To determine the distribution of wildlife in relation to water-holes.
(2) To isolate the relative influences of the vegetation types, seasonality and residual factors on the distribution of wildlife in relation to water-holes.
(3) To make recommendations on the desirability of further development of artificial water-holes in Tsavo National Park (East) Kenya.

The work reported here was carried out in Tsavo National Park, Kenya. Tsavo National Park having an area of 20 000 sq. km, is situated in south-eastern Kenya roughly midway between Nairobi and Mombasa. The Nairobi–Mombasa railway line and tarmac road divide the park into two administrative units — Tsavo National Park West (7500 sq. km) and Tsavo National Park East (12 500 sq. km). Tsavo West is adjacent to the Nkomazi Game Reserve from which it is separated by the Tanzania–Kenya international boundary. During the study period, reconnaissance surveys were made on the ground and by air over most of the dry 'Nyika' (*Acacia–Commiphora* bush) which covers the entire region but because of limited time and logistics the intensive study was confined within Tsavo East which is south of the Voi river. The study area consisted of a more open portion of the park with abundant natural water-holes and some artificial water-holes.

The events leading up to the creation of the park (Sheldrick, 1973); a brief history of ecological research there (Glover, 1974); descriptions of its topography, geology and hydrology (Butler, 1969; Sanders, 1959; Miller, 1952); and various discussions of aspects of the climate (Tyrell, 1972); vegetation (Napier-Bax and Sheldrick, 1963; Greenway, 1969; Norton-Griffiths, 1972) and animal population changes (Glover, 1963; Glover and Sheldrick, 1964; Laws, 1969, 1970; Glover, 1974; and Corfield, 1973) have been reported.

The study period, April 1973–April 1974, was divided into three seasons based on phenological observations of the vegetation (e.g. greenness) as follows: Season I, early dry season (June–August 1973) and late dry season (September–October 1973); Season II, short rains (November–December 1973); and Season III, early green season (January–February 1974), late green season (April–May 1973 and March–April 1974). The seasonal frequencies of some game per km of park road were compared throughout the study period to describe the effects of seasonality on the distribution of wildlife in the study area.

As no comprehensive vegetation map of Tsavo East was available, in order to obtain some idea about possible habitat preferences of certain species, the road counts were analysed for three categories of structural vegetation types (Greenway, 1969); A—grassland; B—bushland and C—woodland/wooded grassland. This classification is based on the physiognomy or the structural appraisal of the vegetation along the census routes rather than the phytosociology of the vegetation communities over the entire study area (Leuthold and Sale, 1973). Each road unit was also assigned to one of the three vegetation categories.

Vegetation A (grassland)

This is land with grasses and perennial herbs, sometimes with evergreen or deciduous trees or shrubs either very scattered or in small isolated groups, in neither case covering more than 10 per cent of the ground.

Vegetation B (bushland)

More than 50 per cent of this land is covered by shrubs and small trees growing densely together. The bushes may be evergreen and have no clearly defined boles. Herbs are ephemeral and/or succulent and grasses are mostly annuals forming ground cover under deciduous bushland.

Vegetation C (woodland/wooded grassland)

This consists of land with an open cover of trees, their crowns not forming a thickly interlaced canopy except along the fringing forest between Voi and Neololo along the Voi River. Scattered evergreen shrubs are present but not conspicuous. Herbs are perennial or annual; grasses form the ground cover. Trees and shrubs in the wooded grassland cover less than 50 per cent of the ground.

METHODS

Road counts

Counts of game species were made during daylight hours by two observers in a Toyota Land Cruiser. The road counts were made every month from April 1973 to April 1974 except in January 1974. As far as possible a sampling strip 300 m wide on each side of the vehicle was observed. The intensive study area was divided into 40 unequal units identified by permanent features such as road junctions and artificial reservoirs. The frequencies of sighting various species of animals per kilometer of road were calculated for each sampling unit.

At the peak of the long dry season, September to October 1973, the precise locations of dry-season water supplies where drinking-water was available for wildlife within the intensive study area were surveyed and mapped. Concentric circles of 5 km radius from the centre of the water supplies were drawn the distance between dry-season water supplies and each of the road units was recorded.

Aerial counts

There are more artificial water-holes along the tourist routes than elsewhere in the study area so in order to describe an unbiased distribution of wildlife in relation to the water-holes, aerial censuses were carried out. In March, August and September 1973 three aerial censuses were carried out by S. M. Cobb in the intensive study area. During each of the three censuses all the natural water-holes in the area had dried out. Seven other aerial counts were made — one each month from the end of October 1973 to April 1974. In each month the count was made after it had rained sufficiently for some natural water-holes to contain enough water for use by game.

A piper super-cub aeroplane (PA-18) with high wings was used during the aerial census, and the game and water-holes seen were recorded using a tape recorder or dictaphone when they fell within a strip whose projection on the ground was 300 m wide. A counter was used when groups of animals were seen together. Systematic flight lines, 5 km apart, were maintained throughout the counts. The counted strip was demarcated by two streamers attached to the strut of one wing.

From the results of the seven wet months and three dry months aerial censuses, distribution maps of game species were constructed by dividing the total number of animals seen in all counts in each 25 sq. km grid within the intensive study area by the number of times (or months) that the square was counted.

The aerial and ground censuses were undertaken to provide quantitative evidence on the distribution of wildlife in relation to water-holes in the intensive study area. It was necessary to embark on both aerial and ground counts because the larger species (elephant and buffalo) which move about in large, closely packed herds can only be counted reliably by air while the smaller species (Peter's gazelle, wart-hog etc.) are more difficult to sight from the air than from the ground. Also without recourse to driving off the park roads (against park regulations), the distribution and the conditions of the water-holes, away from the census routes, can only be assessed from the air.

Analysis

In considering differences between the frequencies of observation of animals in areas (vegetation types or various distances from the dry-season water supplies) of unequal size the number (X) observed in the smaller area (A) and the number (Y) in the larger area (B) were compared with the number that would be expected if the animals had been distributed at random over the intensive study area. Thus, if $X+Y = N$ and $A+B = Z$, there would be

$(A/Z)N$ animals in A and $(B/Z)N$ animals in B. A comparison was then made of X with $(A/Z)N$ and Y with $(B/Z)N$, the observed and the expected values respectively. The significance of the difference was determined by d-test using the ratios of the proportions of observation of game and those of expected values based on random distribution for $n = 30$.

The Kendall rank correlation coefficient was calculated in order to evaluate the degree of association between frequencies of species (X), vegetation types (y) distances from dry-season water supplies (z), and seasonality (W). The coefficients of correlation were generalized to partial correlation coefficient in order to separate out the relative influences of the vegetation types and the distances from the dry-season water supplies on the wildlife frequencies.

In order to decipher which of the variables (x,z,w) account most for the observed frequencies, the d-test was again carried out to determine the significance of the difference between the frequencies of game species within various vegetation types, at various distances to water and for different seasons. Simple non-parametric correlation coefficients (r_{xy}, r_{yx}) and partial correlation coefficient $(r_{xy.z}, r_{xz.y})$ were calculated in order to separate the influences of vegetation types and distances to drinking-water on the frequencies of game species.

RESULTS

Tables 1 and 2 give the lengths of the park roads and the relative proportions of the three categories of structural vegetation types A, B and C at various distances from the dry-season water supplies. Twenty-five per cent of the surveyed road lengths were 0–5 km from dry-season water supplies and 42 per

TABLE 1

Distribution of sampled area in relation to vegetation types and distance from dry-season water supplies (DSWS).

Distance from DSWS	Park roads (km)	% of park roads	Vegetation types (km)		
			A	B	C
0–5 km	64	25	39	20	5
5–10 km	81	33	40	32	9
10 km	104	42	55	6	43
Total km	249	100	135	58	57
			54%	23%	23%

A – Grassland.
B – Bushland.
C – Woodland/wooded grassland.

TABLE 2

Breakdown of sampled vegetation indicating relations between vegetation types and distance from dry-season water supplies (DSWS).

Distance from DSWS	Proportion by distance of vegetation types				Distribution of vegetation types		
	A	B	C	Total	A	B	C
0–5 km	61	31	8	100%	29	35	9
5–10 km	40	40	11	100%	30	55	16
10 km	53	6	41	100%	41	10	75
					100%	100%	100%

A – Grassland.
B – Bushland.
C – Woodland/wooded grassland.

cent of the roads were at 10 km from them. A greater percentage of the area was open grassland, vegetation type A constituting 54 per cent of the entire surveyed area. Along the census routes, however, vegetation type A predominated (75 per cent) at 10 km from the dry-season water supplies but vegetation type B predominated (55 per cent) at 5–10 km from them.

The frequencies of game along the park roads

The mean annual frequencies of observations of some animals within a 600 m belt along monthly census routes, mainly south of the Voi River, are presented in Table 3. The mean frequencies of sighting game per km of census routes varied from 0.070 (giraffe) to 1.127 (zebra). This shows that the zebra was the most common species seen along the survey routes in the intensive study area.

Distribution of surface water and game species by seasons

During the long rains (November–December) and the short rains (April–May) many natural water-holes are filled with rain-water and drinking-water is available within 2 km of all animals. The natural water-holes dry out during the long dry season (June–October). Between rainfall and the number of water-holes containing water between August 1973 and April 1974 when aerial surveys of the number and conditions of the water-holes were carried out in the intensive study area, there was, as expected, a positively significant correlation ($r = +0.810, P = 0.01$).

Although the values of potential evapotranspiration could not be measured during the study period, long-term average monthly values for the Voi area of Kenya are available based on data for 1938–62 and 1964–66 using Penman's

TABLE 3

Mean annual numbers of game per km of park road at Tsavo East

Species	Mean ± S.D.	Remarks
Zebra	1.127±0.604	throughout the study area
Elephant	1.103±0.351	,, ,, ,, ,,
Impala	0.533±0.194	,, ,, ,, ,,
Kongoni	0.480±0.144	,, ,, ,, ,,
Oryx	0.453±0.130	,, ,, ,, ,,
Peter's gazelle	0.303±0.172	,, ,, ,, ,,
Wart-hog	0.197±0.136	,, ,, ,, ,,
Ostrich	0.080±0.044	,, ,, ,, ,,
Giraffe	0.070±0.056	,, ,, ,, ,,
Water-buck	0.513±0.440	along the river
Buffalo	0.333±0.108	near permanent water-holes

formula (Penman, 1948). Figure 3 shows that more game visit Aruba during the dry season when the index of aridity is high than during the rains. There is a positive correlation between the average index of aridity (potential evapotranspiration − precipitation) and intensity the use of water-holes.

Distribution of animal species in relation to the water-holes and the vegetation types

The results derived from road censuses, Table 4, show that in the dry season water-buck, zebra, buffalo etc. were more frequent around the dry-season water supplies from which they dispersed during the rains. The frequency of sighting game species within 0–5 km of dry-season water supplies was significantly higher ($P = 0.001$) than the expected values for the April and October ground counts. In November and December after it had rained, animals moved away from the artificial water supplies and were randomly distributed throughout the study area especially at distances 10 km away from the dry-season water supplies where their frequencies had previously been significantly lower ($P = 0.001$) than the expected values.

Table 5 shows the mean percentages of various species population in the intensive study area which consisted of three categories of vegetation (grassland, bushland and woodland/wooded grassland (Greenway, 1969)). The results show that the numbers of individuals of the species adapted to live in open plains (grazers: e.g. hartebeest, zebra, Peter's gazelle and ostrich) were significantly higher than the expected values in the grassland, whereas the numbers of individuals of browsers (e.g. giraffe) were higher than the expected values in the woodland/wooded grassland during the green and the dry seasons.

The oryx prefers the grasslands in the green season, moving into the wooded grasslands during the dry season. Although the elephant was randomly

TABLE 4

Average seasonal density of wildlife (per sq. km) at various distances from the water-holes containing water along the park roads during the dry season

Species	Distance to water (km)	Seasons		
		I	II	III
Zebra	0–5	48(+)	23	39
	5–10	32	26	25
	10	20(−)	51	36
Elephant	0–5	55(+)	31	44(+)
	5–10	36	33	24
	10	9(−)	36	32
Impala	0–5	56(+)	31	36
	5–10	34	11(−)	12(−)
	10	10(−)	58	52
Kongoni	0–5	45(+)	21	52(+)
	5–10	26	39	23
	10	29	40	25
Oryx	0–5	32	20	41(+)
	5–10	30	28	24
	10	38	52	35
Peter's gazelle	0–5	55(+)	35	36
	5–10	17(−)	34	30
	10	28	31	34
Wart-hog	0–5	78(+)	3(−)	64(+)
	5–10	12(−)	17(−)	25
	10	10(−)	80(+)	11(−)
Ostrich	0–5	68(+)	25	48(+)
	5–10	16(−)	10(−)	20
	10	16(−)	65(+)	32
Giraffe	0–5	81(+)	90(+)	78(+)
	5–10	19	10(−)	16(−)
	10	0(−)	0(−)	5(−)
Water-buck	0–5	85(+)	83(+)	87(+)
	5–10	14(−)	16(−)	13(−)
	10	1(−)	1(−)	0(−)
Buffalo	0–5	65(+)	0(−)	36
	5–10	35	0(−)	64(+)
	10	0(−)	0(−)	0(−)

Seasons: I–dry season (June–October).
　　　　 II–short rains (November–December).
　　　　 III–green season (January–May).

TABLE 5

The seasonal percentages of species population and the significance of the differences from random distribution in vegetation type

	Seasons	Vegetation Types		
		A	B	C
Zebra	Green season	70(+)	9(−)	21
	Short rains	65	20	15
	Dry season	77(+)	20	3(−)
Elephant	Green season	78(+)	8(−)	14
	Short rains	64	7(−)	29
	Dry seasons	61	20	19
Impala	Green season	56	10(−)	34(+)
	Short rains	32(−)	14	54(+)
	Dry season	50	14	36(+)
Kongoni	Green season	71(+)	16	13
	Short rains	69(+)	10(−)	21
	Dry season	69(+)	15	16
Oryx	Green season	69(+)	19	12
	Short rains	50	12(−)	38(+)
	Dry season	57	10(−)	33(+)
Peter's gazelle	Green season	79(+)	12	9
	Short rains	73(+)	9(−)	18
	Dry season	77(+)	13	10
Wart-hog	Green season	20(−)	63(+)	17
	Short rains	79(+)	11(−)	10
	Dry season	43(−)	32	25
Ostrich	Green season	89(+)	8(−)	3(−)
	Short rains	47	28	25
	Dry season	74(+)	16	10
Giraffe	Green season	25(−)	13	62(+)
	Short rains	21(−)	6(−)	73(+)
	Dry season	3(−)	10(−)	87(+)
Water-buck	Green season	37(−)	47(+)	16
	Short rains	57	41(+)	2(−)
	Dry season	17(−)	54(+)	29

(Percentages expected on the basis of random distribution): A = 54%, B = 23%, C = 23%.
*Where using d-test observed values differ from expected at P 0.001, this is indicated.
Seasons: Green season–April–May 1973, January–April 1974.
  Short rains–November–December 1973.
  Dry season–June–October 1973.

A–Grassland.
B–Bushland.
C–Woodland/wooded grassland.

75

distributed in the woodland/wooded grassland throughout the year its frequency was higher than expected in the grassland in the green season. This was consistent with the catholic feeding habits of the elephant; i.e. grazing during the green season and browsing in the dry season.

The water-buck and wart-hog show a preference for bushland. The water-buck was frequently encountered around Aruba Dam where the bush had been mechanically reduced. The restricted distribution of water-buck throughout the year in the vicinity of drinking-water (see Table 4) was consistent with its frequency which was high in all seasons in the bushland associated with the Aruba Dam.

Giraffe and impala show a preference for the woodland/wooded grassland. Giraffe is a browser preferring the woodland/wooded grassland near the Voi River (Table 4) throughout the year and impala is a grazer preferring the wooded grassland along old watercourses.

Aerial census

Table 6 shows the aerial maps and the explanation of the distribution of 11 herbivorous species during the dry and the rainy seasons in the intensive study area. In each case the top figure (a) represents the dry season distribution while the lower figure (b) represents species distribution in the rainy seasons.

During the dry season zebra were often observed west of the study area; at the break-pressure tank near Maungu, near the Aruba dam and Ndara borehole. During the wet season zebra were more frequently observed east of the study area and in larger herds than during the dry season. Elephants were commonly seen within 0–10 km of the Voi River between Voi Gate and Aruba

TABLE 6

Abundance categories for eleven species in Tsavo East south of the Voi River

Species	Abundance categories				
Zebra	(1)	(2)	(3)	(4)	(5)
Zebra	10	5–10	3–5	1–3	1
Elephant	25	15–25	10–15	3–10	3
Impala	5	1–5	1		
Hartebeest	2	1–2	1		
Oryx	5	1–5	0.5–1	0.5	
Peter's gazelle	5	1–5	0.5–1	0.5	
Wart-hog	1–5	0.5–1	0.5		
Ostrich	1	0.5–1	0.5		
Rhino	0.3–0.5	0.1–03	0.1		
Giraffe	1–5	0.5–1	0.5		
Eland	1	0.5–1	0.5		

Entries are of number of individuals per 25 sq. km.

Dam during the dry season. During the rains isolated pockets of well-packed herds of elephant are encountered — the rest being uniformly distributed throughout the study area. The distribution of zebra, elephant and rhino appeared to have shifted from the west to around dry-season water supplies during the rains.

It is more difficult to relate the distribution of the other species — impala, hartebeest, oryx, Peter's gazelle, wart-hog, ostrich, giraffe and eland — to the dry-season water supplies from the aerial counts. The species are in many cases fairly common 5 km from dry-season water supplies during the dry season. During the rains the species move in all directions slightly away from their dry-season distribution areas.

Degree of association between wildlife and some factors of the habitat

Table 7 shows the correlation coefficients r_{xy}, r_{xw} and the partial correlation coefficients $r_{xy.z}$ and $r_{xz.y}$ for 11 herbivores. There is a significantly high positive correlation between the density of giraffe and vegetation types and a negative correlation between the densities of zebra, Peter's gazelle, kongoni and ostrich. The giraffe — a browser — increases in frequency from vegetation type A to C while ostrich, kongoni, Peter's gazelle and zebra — grazers — decrease in density from vegetation types A to C.

There is also a positive correlation between the density of oryx and distance

TABLE 7

Relations of species frequencies with habitat factors as indicated by Kendall's rank correlation coefficient and by partial correlation analysis

Species	Coefficients				
	r_{xy}	r_{xz}	r_{xw}	$r_{xy.z}$	$r_{xz.y}$
Oryx	−0.078	+0.380	−0.067	−0.143	+0.374
Buffalo	+0.172	−0.778	+0.552	+0.101	−0.768
Water-buck	−0.062	−0.860	−0.067	−0.361	−0.879
Wart-hog	−0.174	−0.426	−0.345	−0.261	−0.462
Giraffe	+0.593	−0.773	+0.207	+0.770	−0.866
Kongoni	−0.343	−0.265	+0.600	−0.398	−0.337
Peter's gazelle	−0.312	−0.164	+0.207	−0.343	−0.221
Ostrich	−0.424	−0.237	+0.207	−0.480	−0.331
Zebra	−0.343	−0.140	+0.067	−0.370	−0.202
Elephant	−0.250	−0.385	+0.467	−0.550	−0.438
Impala	+0.157	−0.094	+0.333	+0.146	−0.074

$r_{yz} = 0.14$

x—frequency per km.
y—vegetation types.
z—distances from dry-season water supplies.
w—seasons.

from dry-season water supplies, the frequency of oryx, an arid adapted species, increases away from them. On the other hand the frequencies of giraffe, buffalo, water-buck, elephant and wart-hog, the less arid-adapted species, decrease away from dry-season water supplies.

The association between species frequency and vegetation types, separated from the effects of distance from dry-season water supplies, gives a higher partial coefficient of correlation in the case of ostrich, kongoni, Peter's gazelle, elephant and zebra than the association between the numbers of individuals of these species and distance from supplies. The corresponding association between the species frequency and distance to dry-season water supplies, is higher however, for giraffe, buffalo, water-buck, oryx and wart-hog than the association for these between frequency and vegetation types.

Table 8 summarizes the results of the analysis of partial correlation coefficient for the eleven herbivore species. The regular drinkers, buffalo, water-buck and wart-hog increase in frequency near dry-season water supplies whereas the irregular drinker, oryx, decreases in frequency near supplies. The correlation between the frequency of giraffe in relation to vegetation types and

TABLE 8

Summary of the results of the Kendall rank correlation analysis

Species influenced more by distance from water	
Species increasing away from the DSWS	Species decreasing away from the DSWS
Oryx	Buffalo
	Water-buck
	Wart-hog
Species influenced more by vegetation types	
Species increasing from vegetation A to C	Species increasing from vegetation C to A
Giraffe	Kongoni
	Peter's gazelle
	Ostrich
	Zebra
Species influenced by both vegetation and distance from water	
Elephant	
Species influenced by residual factors and factors other than those above	
Impala	

A—Grassland, B—Bushland, C—Woodland/wooded grassland.

distance to dry-season water supplies are very high, reflecting the giraffe's preference for the riverine zone of vegetation type C. The correlation of frequency with vegetation types and distance to drinking-water are very low in the case of impala showing that residual, and factors other than those of distance to water and vegetation type influence the distribution of the animal.

DISCUSSION

During the dry season the natural water-holes and seasonal rivers were seen to dry out and the vegetation, especially the grasses, to dry up. Thus seasonality affected the quality and quantity of food, the spatial distribution of game biomass and the availability of drinking-water in the various vegetation types. A residual factor, which also affected the frequency of wildlife, was the visibility of various sizes of animals within the sampling strip. The larger species were more easily seen from a moving vehicle in the woodier vegetation types whereas the smaller species were less easily sighted. The nocturnal habits of some species and the periodicity of censuses also influence the frequencies of species of game during the road counts. In the dry season for instance, rhino might not be sighted at all throughout a month due to the intense heat from which they sought shade at midday but during the rains they were frequently recorded in both aerial and road counts and appeared to become more active during daylight hours. The carnivores, on the other hand, were more active at night (when they were often encountered at the water-holes) than during the day.

The kongoni appeared to be an 'obligate' drinker. Ayeni (1975a) suggested a somewhat opportunistic tendency, that is, an aggregation of kongoni near water when drinking water-supplies were less than 5 km away and a distribution influenced by the vegetation types when water supplies were further away. These findings agree with the results of Squires and Wilson (1971) who demonstrated that as the distance between food and water increased, the water and food intake of merino and Border Leicester sheep decreased in the semi-arid areas of Australia. Thus it would appear that very little advantage were served by developing water-holes for the arid-adapted grazers — kongoni, ostrich, Peter's gazelle, impala, oryx, gerenuk and lesser kudu in Tsavo East.

The frequencies of water-buck and giraffe were consistently higher than expected ($P = 0.05$) near the permanent water supplies. The high physiological requirement of water-buck is consistent with its restricted distribution near water. Giraffe, on the other hand, drink sparingly preferring the riparian forest and the riverine vegetation types. Thus the frequent occurrence of giraffe near water was to obtain browse from the trees along the river.

The moderately high coefficient of partial correlation values in Table 6 showed that both vegetation types and distance to water were important co-determinants of frequencies of elephant. This is consistent with the catholic/voracious feeding habits of the elephant (grazing mainly in wet season and browsing primarily in the dry season) and the fact that elephants, especially nursing mothers and their calves, frequent water more regularly in the dry season. The numbers of buffalo were consistently higher than expected ($P = 0.05$) close to water in the dry season the converse being the case with respect to oryx which is an arid-adapted species. Certain growth stages of short grasses attract zebra which aggregate around the water-holes sustaining such growth stages of grasses in the dry season. Thus it appeared that there was scope to primarily attract certain wildlife — elephant, rhinoceros, buffalo, water-buck and possibly zebra to other parts of the park far away from the existing dry-season water supplies by providing drinking-water in such places.

The desirability of management decisions in Tsavo East to create additional water-holes can be better evaluated by considering the history of the park. The major ecological problem in Tsavo East is the elephant population which is considered to be higher than the environment can support leading to widespread destruction of the original woodland. Until 1969, when fire-breaks were constructed along the park boundaries, the reduction of woody vegetation cover continued to be pronounced along the western, southern, and eastern boundaries of the park. In these areas accidental fires originating outside the park (and from honey poachers inside the park) often penetrated into the park and retarded the regeneration of woody vegetation. So, in Tsavo East, the original *Commiphora* woodland is being degraded by elephants and until recently by fire, into varying degrees of open and semi-open mosaics of *Boscia Platycelyphium-Sericocomopsis* wooded grassland. One of the initial effects of the change was believed to be an increase in the population of grassland-adapted species such as zebra, oryx and kongoni, whereas woodland-adapted species such as rhinoceros and lesser kudu were thought to have decreased (Glover, 1963; Napier-Bax and Sheldrick, 1963; Agnew, 1968; Laws, 1969; Glover, 1974).

During the 1961 drought in Eastern Kenya, the death of about 300 rhinos in Tsavo East was believed to have been caused by the comparative inability of rhino to compete with elephant for browse and shade (Napier-Bax and Sheldrick, 1963; Glover and Sheldrick, 1964). Laws (1970) suggested that the only feasible solution to the elephant problem, albeit repugnant, was the reduction of elephant populations through cropping programmes.

Although there has been much concern over the consequences of Tsavo East elephant-induced habitat change, a suggested work-plan for the management of elephants based on systematic cropping led only to further controversy. During the 1970–71 drought, another high mortality, this time of some 6000 elephants occurred in the park mainly in Tsavo East. Corfield (1973) suggested that the differential nature of the mortality pattern, selectively removing the reproductive females and young animals, improved the chance of survival of existing population of elephants in Tsavo East, in the absence of immigration. Thus cropping may not be the only feasible way of reducing the number of elephant to within the carrying capacity of the habitat.

Immigration and emigration might have been important in influencing elephant numbers and movement in Tsavo in the past but their significance has not been quantitatively studied in Tsavo East. Substantial movement of elephants occurs and may explain, in part, that in spite of the high mortality of elephants in Tsavo, subsequent estimates have shown that numbers remained unchanged from 1970–71 population of 14 500 in Tsavo East (Leuthold and Cobb, personal communications).

Elephants which constitute two-thirds of total game biomass, and other game in Tsavo East were thought to have moved near the permanent river in the dry season (Glover, 1963) so additional water supplies were developed in order to spread the big game more evenly, away from the Galana River. During prolonged drought, the aggregation of game on the narrow riverine belts resulted in over-exploitation of the habitat and ultimately a high game-mortality — as was the case in 1960–61 and 1970–71 when many rhinos and elephants perished in Tsavo East.

Although it is apparent that with more water-holes which contained water in the park, more large game species — rhino, elephant and buffalo — would use the park, it is not clear whether such a change would be desirable. An increase in the number of elephants, the largest ungulate in Tsavo East, could well lead to a further destruction of woody vegetation, an effect which has already reached deplorable proportions in the park. Also elephants and rhinos are known to have died in the park more as a result of starvation rather than thirst since most of the carcasses were found near permanent water supplies. If animals could live solely on drinking-water rather than food (vegetable diet) during a drought then the mortalities of rhino and elephant, could have been averted by the development of water-holes but, obviously, this is not the case.

The provision of water-holes did not, in any way, result in an improvement of food quality around the drinking-water supplies and as such could not stop game mortality caused by starvation.

However, it can also be argued that if additional water-holes were not developed in Tsavo East, wildlife would migrate into adjacent ranches. The park is surrounded by ranches where major water-development projects are being undertaken to provide water for livestock. It was thought that a corresponding development of additional water-holes within the park was required in order to counteract the effects of the water development programmes outside the park. Here, the quantitative studies of game movements necessary to ascertain whether any game movement from the park into areas outside it was directly related to the creation of additional water supplies on the ranches were not available. From the present study, and other preliminary results on radio telemetry in the dry season, elephants are found to have relatively restricted home-ranges near permanent water-holes from which they disperse, during rains, drinking from temporary water-holes formed in clay pans filled with rain-water (Leuthold and Sale, 1973). Thus this alternative argument for the creation of water-holes is not tenable in the light of present understanding — but further investigation would be desirable.

The following tentative conclusions and recommendations can be made from this study:

(1) Although at present the development of water-holes on the ranches poses no threat as a possible factor attracting wildlife from the park in the dry season, the land-use outside the park should be monitored continuously to predict more accurately when it may influence the ecology within the wildlife sanctuary.

(2) The creation of additional large permanent water-holes (e.g. Aruba Dam) will, over the short term, be reflected in increases in the elephant population through reduction of 'drought-linked' mortalities and further aggravate the reduction of woody vegetation cover, in the long run, making it ultimately necessary to regulate elephant number artificially. Sub-surface weirs as preferable alternatives to the dams could be constructed along the rivers. During the dry season, water would be trapped in the silt which elephant can dig out thus opening up drinking-water supplies for many more wildlife species.

(3) The number of elephants which the riverine vegetation can support during the dry season should set the upper limit to the desirable number of elephants in the park. Excess elephant biomass above the carrying capacity along the rivers would die in subsequent dry seasons until a steady state of equilibrium is reached in the elephant population.

(4) In other African national parks with higher rainfall and dense forests, where habitats have not been degraded by a high large-mammal biomass the judicious development of water-holes should only commence after ensuring that:

(a) the quality of the water supplied to game would not adversely affect animal health and habits;

(b) there are no serious risks of encouraging pests and disease transference through increased water-hole; and

(c) the creation of the water-holes does not adversely alter spatial distribution of animal species and vegetation types, and that while enhancing game viewing, the micro-flora and micro-fauna in the park are not disturbed.

REFERENCES

A.R.C. (1965) The nutrient requirements of farm livestock. *Technical Reviews and Summaries, Ruminants.* No. 2 (Agricultural Research Council, London).

AGNEW, A. D. Q. (1968) Observations on the changing vegetation of Tsavo National Park (East). *East African Wildlife Journal*, pp. 78–80.

AYENI, J. S. O. (1975a) Periodicity of African wildlife at water-holes in Tsavo National Park (East), Kenya. *Bull. Anim. Hlth. Prod, Afr.*, Vol. 23, No. 2.

AYENI, J. S. O. (1975b) Utilization of water-holes in Tsavo National Park (East). *East African Wildlife Journal*, Vol. 13, pp. 305–23.

BUTLER, R. J. Ed. (1969) *Atlas of Kenya*. (The Survey of Kenya, Nairobi).

CORFIELD, T. F. (1973) Elephant mortality in Tsavo National Park, Kenya. *East African Wildlife Journal*, Vol. 11, pp. 339–68.

FINCH, V. A. (1972) The effects of solar radiation, of temperature regulation, and heat balance in two East African antelopes, the eland and the hartebeest. *American Journal of Physiology*, Vol. 222, pp. 1374–9.

GLOVER, P. E. and SHELDRICK, D. L. W. (1964) An urgent research problem on the elephant and rhino populations of the Tsavo National Park in Kenya. *Bull. eqpiz. Dis.*, Vol. 12, pp. 33–8.

GLOVER, P. E. (1970) The Tsavo and the elephants. *Oryx*, Vol. 10, pp. 323–5.

GLOVER, P. E. (1974) A short history and summary of work undertaken by the Tsavo Research Project (Mimeographed report). (Kenya National Parks, Nairobi).

GLOVER, J. (1963). The elephant problem at Tsavo. *East African Wildlife Journal*, Vol. 1, pp. 30–9.

GREENWAY, P. J. (1969). A checklist of plants recorded in Tsavo National Park, East.*J. E. Afr. Histo. Soc.*, Vol. 27, pp. 169–209.

JARMAN, P. J. (1973) The free water intake of impala in relation to the water content of their food. *East African Agricultural and Forestry Journal*, Vol. XXXVIII, No. 4, pp. 343–51.

LAWS, R. M. (1969) The Tsavo Research Project. *Journal of Reproduction and Fertility*, Suppl. 6, pp. 495–531.

LAWS, R. M. (1970) Biology of African Elephants. *Science Progress*, Vol. 58, pp. 251–62.

LEUTHOLD, W. and LEUTHOLD, B. (1973) Ecological studies on ungulates in Tsavo National Park (East), Kenya. Research Project. *(Kenya National Parks, Nairobi)*.

LEUTHOLD, W. and SALE, J. B. (1973) Movements and patterns of habitat utilisation of elephants in Tsavo National Park, Kenya. *East African Wildlife Journal*, Vol. 11, pp. 369–84.

MACFARLANE, M. V., WARD, N. and SIEBERT, B. D. (1967) Water metabolism of Merino and Border Leicester sheep grazing saltbush. *Journal of Agricultural Research*, Vol. 18, pp. 947–58.

MILLER, J. M. (1952) *Geology of Mariakani — Mackinon Road Area*. (Geological Survey of Kenya).

NAPIER-BAX, P. and SHELDRICK, D. L. W. (1963) Some preliminary observations on the food of elephants in Tsavo Royal National Park (East) of Kenya. *East African Wildlife Journal*, Vol. 1, pp. 40–53.

NORTON-GRIFFITHS, M. (1972) Report on the aerial photography carried out in the Tsavo National Parks in June 1972 with recommendation for a park-wide sampling design. *Serengeti Ecological Monitoring Programme* (S.R.I. Arusha, Tanzania).

PENMAN, H. L. (1948) Natural evaporation from open water, bare soil, and grass. *Proceedings of the Royal Society*, Vol. 193, pp. 120–45.

SANDERS, L. D. (1959) Geology of the Mid-Galana area. Ministry of commerce and Industry. *(Geological Survey of Kenya)*.

SHELDRICK, D. (1973). *The Tsavo Story*. (Collins and Harvill Press, London).

SQUIRES, V. R. and WILSON, A. D. (1971). Distance between food and water supply and its effect on drinking frequency, food and water intake of merino and Border Leicester sheep. *Journal of Agricultural Research*, Vol. 22, pp. 283–90.

TAYLOR, C. R., SPINAGE, C. A. and LYMAN, C. P. (1969) Water relations of the water-buck, an East African antelope. *American Journal of Physiology*, Vol. 217, pp. 630–4.

TYRRELL, J. G. (1972) Aspects of water balance of Tsavo East (Research Progress Report). Unpublished.

SANDERS, L. D. (1963). Geology of Voi — South Yatta area. Ministry of Commerce and Industry. *(Geological Survey of Kenya)*.

YOUNG, E. (1970). Die Ekologie van Wild in die Nasional Krugerwildtruin. D.SC. Thesis, Pretoria University.

SECTION 2
HABITAT UTILIZATION

10

BIG GAME UTILIZATION OF NATURAL MINERAL LICKS

J. S. O. Ayeni

Kainji Lake Research Institute, New Bussa, Kwara State, Nigeria

INTRODUCTION

Natural mineral licks are mineral outcrops in the soil which are visited by herbivores for soil eating (biting and chewing) or licking (with the tongue). Ayeni (1971) reviewed some literature on licks from North America and noted that the early authors had emphasized the mineralogical content of licks rather than the frequency and behaviour of animals at the licks.

A detailed study of licks in Yankari Game Reserve in the Bauchi State of Nigeria, was undertaken from the end of the dry season — June 1970, to the middle of the rainy season — August 1970. Henshaw and Ayeni (1971) described the level and periodicity of lick use by big game in Yankari. It was suggested, from the results of chemical analysis of soil samples taken from the inside and the outside of licks, that the licks could supplement the minerals (sodium?) that are deficient in animals' vegetable diet (Ayeni, 1972).

Yankari Game Reserve in Bauchi State is located 193 km east of Jos in the Plateau States of Nigeria. It was established in 1956 and opened for tourism in 1962. It is the most developed game reserve in Nigeria, comprising 2077 sq. km of Sudan and Guinea Savannah vegetation types. The two other Nigerian Game Reserves where observations were also taken are: the Kainji Lake National Game Park (in the section formerly known as the Borgu Game Reserve) in Kwara State and Upper Ogun Game Reserve in Oyo State. The last two reserves have a more humid climate than the Yankari Game Reserve.

In East Africa, Tsavo National Park (East) 'Tsavo East' is located about 338 km south-east of Nairobi, Kenya. It has an area of about 12 500 sq. km and contains both the Masai and Somali fauna types. The Tsavo ecological unit is well known for its great elephant *Loxodonta africana* Blumenmach (over 30 000) and black rhinoceros *Diceros bicornis* Linn. (5000) populations. The other East African game reserves and national parks, where further observations were taken, especially on game species that are either uncommon or do not exist in Nigeria, are: Nairobi National Park, Amboseli Game Reserve and Lake Nakuru National Park in Kenya, Lake Manyara National Park,

Ngorongoro Conservation Area and Serengeti National Park in Tanzania. In all the places listed, except Tsavo East, licks are ubiquitous.

METHODS

Although information contained in this paper was gathered from trips to various parks and game reserves, sufficient time was spent in many of the wildlife sanctuaries to make the observations worthwhile. I was in Yankari for three months in 1971 studying some aspects of big game utilization of natural mineral licks. I also worked in the Borgu section of Kainji Lake National Game Park, meanwhile visiting the Upper Ogun Game Reserve between 1971 and 1972. During the 1972–73 field courses from the University of Nairobi, many field trips were made to the areas of East Africa already referred to. In all these places notes were kept on the level of lick utilization, abundance of licks and seasonal difference in lick utilization. During 1973–74 field work on some aspects of water-hole utilization in Tsavo East (Ayeni, 1975) vegetation, soil, and water samples were collected and analysed by mass spectography through the assistance of the chemists in the National Agricultural Laboratories in Nairobi.

RESULTS

Species utilizing natural mineral licks

The following species of wildlife were observed using licks in Nigeria: elephant (*Loxodonta africana* Blumenbach), buffalo (*Syncerus caffer* Sparman), western hartebeest (*Alcelaphus buselaphus* Pallas), roan antelope (*Hippotragus equinus* Desmarest), water-buck (*Kobus defassa* Ogilby), kob (*Kobus kob* Erxleben), bushbuck (*Tragelaphus scriptus* Pallas), oribi (*Ourebi ourebi* Zimmermann), red-flanked duiker (*Cephalophus rufilatus* Gray), Grimm's duiker (*Sylvicapra grimmia* Linnaeus), wart-hog (*Phacochoerus aethiopicus* Pallas), and baboon (*Papio anubis* Fischer). The following other game species were also observed in East Africa to use licks in different places: rhino (*Diceros bicornis* Linnaeus), zebra (*Equus burchelli böhmi* Matschie, and *E. grevyi* Oustalet), impala (*Aepyceros melampus* Lichtenstein), water-buck (*Kobus ellipsiprymnus* Ogilby), eland (*Taurotragus oryx* Pallas), wildebeest (*Connochaetes taurinea* Burchell), giraffe (*Giraffa camelopardalis* Linnaeus), kongoni (*Alcelaphus buselaphus cokei* Günther), topi (*Damaliscus korrigum* Ogilby), Grant's gazelle (*Gazella granti petersi* Günther), and Thomson's gazelle (*Gazella thomsonii* Günther). Some cattle (in Nigeria) sheep and goats (in Kenya at the Suswa Hills) in free range were also observed using licks. No carnivore had been observed to use licks in either East or West Africa. This also appeared to be the case in southern Africa.

The significance of the abundance of natural mineral licks

From experience in the game reserves that are situated in the drier habitats of northern Nigeria, Henshaw and Ayeni (1971) postulated that an abundance of, and an increased use of licks by wildlife would indicate nutritional deficiencies caused by a degrading environment or over-population or both. This appeared to be the case in Yankari where ten licks were within 11 km along the Gaji River.

The frequency of occurrence and the intensity of the use of licks in Yankari was far greater than in Tsavo East. In Tsavo East the climate is extremely dry and the original *Commiphora* woodland is being degraded by elephants, and until recently by fire, into varying degrees of open and semi-open mosaics of *Boscia-Platycelyphium-Sericocomopsis* wooded grassland. Only three licks were seen to attract wildlife in Tsavo East and this happened infrequently (Sheldrick, personal communication). Thus an abundant occurrence of licks is not necessarily a proof of habitat deterioration and/or over-population.

Environmental conditions

The geology of Tsavo (Sanders, 1963) is not very different from that of Yankari (Carter *et al.*, 1963) however the minerals available to vegetation in the two regions may be a result of their respective climates. Yankari receives about 1000 mm of annual rainfall, but Tsavo rarely receives more than 500 mm and has a probability of less than 250 mm annual rainfall once every 10 years. The pattern of rainfall distribution is also different. In Tsavo, there are two rainy seasons in which rain falls in a series of showers spread over many months, but in Yankari the whole annual precipitation falls in a single short rainy season of 2–3 months.

The geology and climate of Tsavo compared with Yankari suggests that the water-soluble sodium salts in Yankari could be leached out during the heavy short rains following the long period of dessication. It can be assumed, therefore, that sodium ions would not be available in the vegetable diet of the game in Yankari since plant species normally absorb only a small fraction of the 'non-essential' sodium ions from the soil. Plants even substitute potassium ions for sodium ions uptake from the soil without showing mineral deficiency symptoms (Buckman and Brady, 1960).

It has been suggested that the shortage (assumed?) of sodium ions in the plants which are eaten by wildlife could motivate big game to eat a lot of soil at the lick (Ayeni, 1972). This assumption is substantiated in the observation of Dougall *et al.* (1964) who noted that the abundance of and the intensive utilization of licks in many parks in Kenya could be due to the fact that the dry-weight composition of Kenya browse and pasture herbage averaged below 0.15 per cent sodium content.

Weir (1972) demonstrated a positive correlation between the spatial distribution of elephants in Wankie National Park, Central Africa, and the abundance of environmental sodium. If this relationship is applicable to Tsavo East, one would expect that sodium ions are widely abundant, since elephants were found almost everywhere during my studies in Tsavo East. The abundance of alternative sources of environmental sodium (see Table 1) made it unnecessary for wildlife in Tsavo East to seek out licks. During the rainy season, wildlife in Tsavo East drank from natural pans, which are formed from termite mounds, where many salts had been precipitated out through evaporation (Hesse, 1955; Watson, 1962; Weir, 1960, 1973). In the dry season, wildlife in Tsavo East also drank from the rivers, dam, reservoirs and boreholes all of which are very rich sources of sodium. Many plants in Tsavo East are also rich in sodium and potassium, for example: *Abutilon mauritianum* (Jacq.), *Acacia elatior* (Brenan), *Achyranthes aspera* (L.), *Calyptotheca taitensis* (Pax Vatke), *Cordia garaph* (Forsk), *Echolium amplexicaule* (Moore), *Premna resinosa* (Hochst) and *Salvadora persica* (L.). The presence of sodium (above 0.15 per cent dry weight) in the plant species listed above shows that Tsavo East is locally rich in environmental sodium.

Table 2 shows the results of chemical analysis of soil samples taken across a catena from the topsoil to the bedrock (Glover, 1970), and how sodium content increased down the profile in Tsavo East. The effect of this is that the water contained in the reservoirs, which were developed by widening and deepening some natural pans in Tsavo East, exhibit higher total sodium content variations in the course of the year than the water contained in the natural pans. The water in the boreholes and near some mineral outcrops in the

TABLE 1

Sodium content of water samples in Tsavo East

Date of sampling the water-holes	Origin of the water samples	Sodium content (ppm)
February 1974	Dida-harea natural pan	65.1
January 1973	Dika natural pan	67.4
February 1974	Kono-moja natural pan	55.0
November 1973	Dida harea reservoir	235.1
August 1973	Dika reservoir	197.6
February 1973	Kono-moja reservoir	265.0
June 1973	Aruba dam	123.8
November 1973	Ndara borehole	1449.9
October 1973	Mukwaju borehole	195.1
January 1974	Tsavo river	102.6
January 1974	Galana river	106.3
November 1973	Voi Safari Lodge (Mzima springs)	279.9
November 1973	Voi river (salt outcrop excavated in riverbed by elephants)	1899.8
January 1973	Buffalo wallows — ,, —	2788.7

TABLE 2

The results of Merlich analysis for sodium (content in ppm) in 12 pits representing 6 different types of soils from which soil samples were collected down the profile in Tsavo East

Pit across the catena	Sodium content of soil samples collected down the profile from top-soil to bedrock							Mean Na+ in ppm	Mean pH in H₂O
	Sample 1	Sample 2	Sample 3	Sample 4	Sample 5	Sample 6	Sample 7		
1	252.89	873.62	2321.99	2482.92	2321.99	4345.11	—	2099.75	8.4
2	91.96	91.96	114.95	137.94	91.96	321.86	390.83	121.53	8.8
3	68.97	68.97	68.97	68.97	114.95	275.88	—	111.11	5.9
4	160.93	68.97	68.97	68.97	91.96	206.91	—	111.11	5.1
5	68.97	91.96	91.96	160.93	160.93	459.80	—	172.42	6.1
6	22.99	22.99	91.96	321.86	919.60	1057.54	—	406.15	7.4
7	91.96	620.73	3034.68	3577.27	2276.01	1586.31	—	1931.16	8.4
8	68.97	22.99	137.94	252.98	367.84	781.66	—	272.05	8.1
9	68.97	45.98	91.96	114.95	495.80	459.80	—	206.91	8.0
10	712.69	22.99	22.99	22.99	68.97	183.92	—	172.42	5.7
11	45.98	68.97	68.97	114.95	114.95	252.89	—	111.11	5.9
12	68.97	68.97	68.97	275.88	459.80	1195.48	—	356.34	6.7

TABLE 3

Periodicity of wildlife to natural mineral licks in Yankari Game Reserve, Nigeria

Time	Hrs of Obs.	Buffalo/hr.	W. Hartebeest/hr.	Water-buck/hr.	Wart-hog/hr.	Baboon/hr.	All species/hr.
01.00	2	—	—	—	—	—	1
02.00	2	—	0.5	—	—	—	1
03.00	4	—	2.0	—	—	—	2
04.00	6	—	0.84	—	—	—	1.3
05.00	9	0.22	0.33	—	—	—	1
06.00	14	—	0.86	—	—	—	1
07.00	15	—	0.73	—	0.4	1.56	2.73
08.00	20	0.15	0.20	—	0.5	1.45	2.30
09.00	31	0.39	0.94	0.13	1.26	2.84	5.62
10.00	35	0.29	0.17	0.29	2.8	1.26	5.84
11.00	35	0.086	1.6	0.4	2.58	0.69	5.35
12.00	40	0.025	0.875	0.98	3.83	1.88	7.63
13.00	36	0.138	1.36	0.87	4.55	1.64	8.83
14.00	33	0.27	2.09	1.67	4.45	0.64	9.13
15.00	31	0.032	1.52	0.77	2.81	0.94	6.10
16.00	24	0.29	1.29	0.167	0.79	0.167	1.92
17.00	15	—	1.00	—	0.33	0.66	1.73
18.00	11	—	0.55	—	0.27	0.55	1.1
19.00	3	—	—	—	—	—	0
20.00	2	—	—	—	—	—	0
21.00	2	—	—	—	—	—	0
22.00	2	—	—	—	—	—	0
23.00	2	—	—	—	—	—	0
24.00	2	—	—	—	—	—	0

All Species = Species above and other species which utilize the licks infrequently.

riverbeds of the seasonal streams, supplement the daily sodium requirement of wildlife species; and as a result the animals aggregate around these water supplies in the dry season.

Activities of some lick-using species

Tables 3 and 4 summarize the periodicity and activities of different lick-using species in Yankari. Although lick use is predominantly a day-time activity, different species used the licks at slightly different periods of the day. Buffalo and hartebeest are partly nocturnal and partly diurnal. Water-buck and wart-hog are entirely diurnal. Baboon are also diurnal and generally show a peak in lick use in the morning and afternoon (see Table 3).

The shapes of the animals mouth parts determine the methods by which the dry-season lick material was obtained and ingested. Elephant and wart-hog dig up the lick contents using their tusks. The elephant lifted up large pieces of the lick material with the trunk, whereas baboon used the hands to pick up and throw small pieces of the material into the mouth. Wart-hog and hartebeest cut fresh lick material by biting deeper into the craters with the incisors, whereas water-buck, buffalo, roan, duiker, oribi and bushbuck lick up the powdered lick-material with the tongue.

In the rainy season, the function of some licks changed to those of natural

TABLE 4

Summary of the activities of wildlife on the lick at Yankari Game Reserve

Species	Toilet	Drink	Lick soil	Cut new soil	Eat loose soil pieces	Dust body	Wallow
Elephant	†			†	†	†	
Buffalo	†	†	†			†	†
Hartebeest	†	†		†		†	
Water-buck	†	†	†				
Wart-hog	†	†		†		†	†
Baboon	†	†	†		†		

Explanation of Table
Toilet   — defecation and/or urination.
Lick   — use tongue to remove powder soil.
Cut new soil   — biting soil.
Eat loose soil — chewing soil.
Dusting   — spray dust on body or rub body against soil.
Wallow   — roll in muddy water or clay.
†   — activity performed.

water-holes from which wildlife drank and wallowed. During the rains, buffalo and water-buck used licks mainly for drinking, but wart-hog and hartebeest still continued to eat lick materials as well as drinking at the licks. In Yankari during the rains wart-hogs often wallowed in the licks, and in East Africa, elephant, buffalo and rhino were also seen to wallow in the natural water-holes formed in zoogenous clay pans.

Avoidance of licks during inclement weather

Although the cumulative number of wildlife increased as the rainy season progressed, due to the use of the licks as water-holes, there was, nevertheless, an inverse correlation between the absolute number of animals visiting the licks and rainfall (see Figure 1). This observation can be explained by the fact that, in the rainy season, water was abundant in natural pans and wildlife dispersed away from the vegetation strips along the river with which the licks are associated and serve as a dry season refuge for game. Most lick-using species also avoided spending the hottest period of the day at the licks where there are very few shade plants (see Table 3) in dry weather. This strategy could assist wildlife to minimize heat load (thermoregulate) at the licks in hot, dry weather.

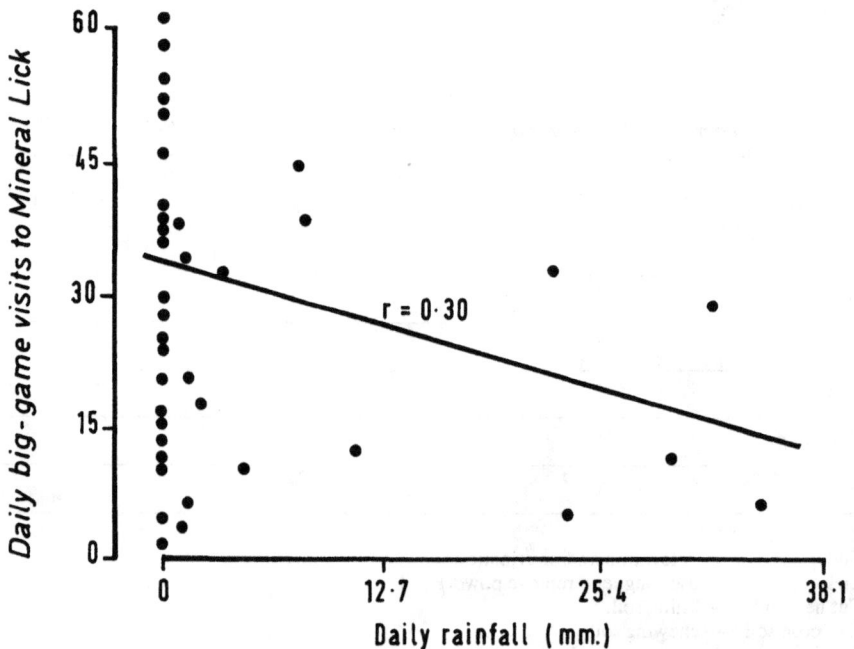

Fig. 1. The relation between lick visiting and rainfall.

Between 14.00 hours and 16.00 hours the average number of animals observed per hour dropped from over eight at the peak before 14.00 hours to less than two (see Table 3). Since the hottest period of the day is usually a little after the maximum insolation phase, it is not surprising that at about 15.00 hours most of the animals at the licks moved away.

DISCUSSION

In many tropical wildlife sanctuaries in Africa wildlife use licks to supplement their sodium requirement, but where environmental sodium is abundant, the soil very saline and the climate arid, the occurrence of licks is very infrequent. This conclusion was recently (March, 1976) confirmed by further observations in the Ruwenzori National Park (formerly known as the Queen Elizabeth National Park) in Uganda. At Ruwenzori, near Mweya and at Shasha, the Uganda kob and other big game were not observed to lick soil; but further north of the park the kobs were seen using the licks at the Semiliki Game Reserve. The difference between the two parks regarding the development and utilization of licks is related to the spatial distribution and abundance of environmental sodium. Ruwenzori soil has a higher sodium content than Semiliki because it is watered by Lake Victoria, indeed there is a table-salt factory at Katwe in Ruwenzori.

Lick use raises three problems, namely: soil removal, vegetation destruction, and spread of diseases. Animals remove over 5000 tons of soil annually from the Treetops licks at Aberdare National Park, Mount Kenya (Woodley, personal communication). The soil is lost through wallowing, eating or licking, and by trampling. The areas immediately around the licks are usually trampled by hoof action, overgrazed, and devoid of vegetation cover. It is necessary to prove whether the plant species that are preferred by game are disappearing around the licks. It may also be necessary to conduct research to ascertain whether it would be desirable and, indeed, possible to make cheap artificial mineral licks for use in areas where certain preferred licks lie outside the boundaries of the game reserves or where suitable soil material cannot be found to replenish the annual soil removal from the licks. Finally, since so much drinking, urination, defecation, wallowing and feeding occur at the licks diseases spread rapidly and artificial licks may prevent this — these aspects should be investigated.

The behaviour of wildlife using the licks, one at a slightly different hour from the others, may serve to minimize interspecific competition though such periodicity of visits may be more related to the daily feeding patterns of different species. The avoidance of licks in the hottest periods of the day, may be related to thermoregulatory mechanism of the game. Lower records of game at particular mineral licks during the rains may be related to the fact that wildlife spread out more and thus utilize a greater area of the park during the rains, drinking from sodium-rich pans.

CONCLUSIONS

Since natural licks concentrate animals in areas of good visibility, they could be of management significance to improve game viewing and photographing for tourists and to adapt as inexpensive areas of conducting animal studies (possibly including censusing) — when hides and tracts are constructed. At Taita Salt Lick Lodge near Tsavo Park and at Treetops (Kenya) for instance, hides for game viewing are being used by tourists for many nights throughout the year.

A possible line of future research would be to examine the desirability of creating artificial licks (where natural ones are few or lacking) and enriching existing natural licks and some water-holes. At present, studies on the impact of lick use on the habitat around the licks and the spread of diseases and parasites at licks should be given highest priority.

Along with the chemical analysis of soils from different game conservation areas, an attempt should always be made to assess the chemistry and distribution of all the plants known to be eaten by the animals. This would make it easy to evaluate and compare the mineral levels available to game species from one area to another. Unlike Tsavo in East Africa, there is very little data available on the chemical composition of Yankari plants and in general such investigations do not rank high on the list of priority projects in West Africa. It would be valuable for such analyses to be given out on contract or on a consultancy basis to university departments or government research projects in West Africa.

ACKNOWLEDGEMENTS

I am grateful to the chemists of the National Agricultural Laboratory, Nairobi where the analysis of the soil, water and plant samples were carried out. Dr P. E. Glover, Biologist Tsavo Research Project and Mr D. L. W. Sheldrick, Senior Park Warden, Tsavo East supplied some information on previous soil analysis in the park. Professor Anthony Youdeowei, Department of Agricultural Biology, University of Ibadan read through the manuscript making useful amendments.

REFERENCES

AYENI, J. S. O. (1971) Natural mineral licks: literature review. *Obeche.* Vol. 7, No. 1, pp. 47–53.

AYENI, J. S. O. (1972) Chemical analysis of some soil samples from the natural licks of Yankari Game Reserve, North-Eastern State, Nigeria. *Nigerian Journal of Forestry*, Vol. 2, No. 1, pp. 16–21.

AYENI, J. S. O. (1975) Utilization of water-holes in Tsavo National Park (East). *East African Wildlife Journal*, Vol. 13, pp. 305–23.

BUCKMAN, H. O. and BRADY, N. C. (1960) Reaction of Saline and Alkali Soil. In *The Nature and Properties of Soils*. 6th Ed. (The Macmillan Company, New York), 567pp.

CARTER, J. D. W., BARABA, E. A., TRAIT, and JONES, G. B. (1963) The Geology of parts of Adamawa, Bauchi and Bornu Provinces in North-East Nigeria. Bull. Geological Survey of Nigeria. No. 30.

DOUGALL, H. Q., DRYSDALE, V. M. and GLOVER, P. E. (1964) The chemical composition of Kenya browse and pasture herbage. *East African Wildlife Journal*, Vol. 2, pp. 86–121.

GLOVER, P. E. (1970) Tsavo Research Project Progress Report (June 1968–June 1970) Kenya National Parks.

HENSHAW, J. and AYENI, J. (1971) Some aspects of big game utilization of mineral licks in Yankari Game Reserve, Nigeria, *East African Wildlife Journal*, Vol. 9, pp. 73–82.

HESSE, P. R. (1955) A chemical study of the soils of termite mounds in East Africa. *Journal of Ecology*, Vol. 43, pp. 449–61.

SANDERS, L. D. (1963) Geology of Voi-South Yatta Area. Report No. 54, (Geological Survey of Kenya).

WATSON, J. P. (1962) The soil below a termite mound. *Journal of Soil Science*, Vol. 13, pp. 46–51.

WEIR, J. S. (1960) A possible course of evolution of animal drinking holes (pans) and reflected changes in their biology. (*Trans. First Fed. Sci. Cong.* Salisbury).

WEIR, J. S. (1972) Spatial distribution of elephants in an African National Park in relation to environmental sodium. *Oikos*, Vol. 23, pp. 1–13.

WEIR, J. S. (1973) Air flow, evaporation and mineral accumulation in mounds of Macrotermes subhyalinus (Rambur). *Journal of Animal Ecology*, Vol. 42, pp. 509–20.

11

HABITAT MANAGEMENT FOR WILDLIFE CONSERVATION WITH RESPECT TO FIRE, SALT LICK AND WATER REGIME

Abdul H. Lasan

Game Preservation Unit, Forestry Division,
Ministry of Natural Resources,
Nigeria

INTRODUCTION

Habitat management must be carried out before any direct management of wildlife. At times, management may consist of total habitat protection to keep it suitable for certain kinds of game, for example the browsers. Under other circumstances drastic modification of the habitat, for example by the use of fire, must be effected if wildlife populations are to be maintained (Dassman, 1964). Among wildlife species the vertebrates have complex nutritional requirements and a number of different chemical elements are needed, e.g. calcium and phosphorus which may be obtained naturally from salt licks. Water supply is also of vital importance.

These three management tools, fire, salt licks and water supply are among the basic ecological requirements in any wildlife conservation area. In this paper each is examined separately with respect to its importance in habitat management for wildlife conservation in East and West Africa.

ROLE OF FIRE

In the early days, man probably used fire first for heating. Presumably he later used it for cooking and in gathering honey. He perhaps soon learned to use it on vegetation to help him hunt and in association with shifting agriculture, felling and burning forest trees and growing crops in the ashes (Barlett, 1957).

Subsequent inhabitants of the tropics and subtropics adopted and perfected early man's method of using fire. Fire is now used to produce the grasslands and savannahs required for pastoralism.

The scientists and managers of game reserves at present advocate using fire as a tool for achieving appropriate integration of the renewable natural resources in their charge. Thus in East Africa, at the Serengeti National Park, comprehensive fire usage is designed for management of the habitat. Some areas are burned early to open up spaces for the plains animals while other areas, usually the thickly vegetated woodland, are burned late to eliminate some seedlings and enable others to grow quicker. Fire is kept away from other areas completely, to accommodate some species of wildlife that need complete cover and to protect the visually attractive riverine vegetation. In Nigeria, at Yankari Game Reserve, some areas are burned early to ensure that tall grasses do not prevent the tourists viewing the game. This early burning also eliminates the possibility of late hot fires. Also, new shoots of fresh grass are encouraged by the early burning. In some areas, however, prevention of burning is enforced to conserve riverine vegetation and to discourage soil erosion.

The value of adequately controlled usage of fire in habitat management is clear from experiences in areas where definite burning policies have been developed.

The effect of regulated burning on the grassland at Serengeti is that some undesirable grasses are ousted by palatable grasses. Thus *Pennisetum mezianum* is discouraged by fire while the desirable *Themeda, triandra*, which is fire tolerant, is encouraged by burning. New shoots of grass that are generally preferred by the animals (Dougall and Glover, 1964) are also encouraged by the regulated use of fire on the grassland — Mess (1958) observed that the feeding value of the new growth on burned grassland is actually higher than that of new growth from unburned grassland.

In the Serengeti woodlands burning is regulated to manipulate tree regeneration which in turn gives suitable habitat for different wildlife species. Regular burning stunts the growth of trees because the young delicate seedlings are burned down. Cessation of burning for a stipulated period will help restore normal growth rate. On the other hand habitats of browsers and riverine areas are totally protected from burning. This encourages thick vegetation and good cover for animals frequenting such vegetation.

In Yankari Game Reserve coarse grasses are eliminated and in their place fresh shoots of both grasses and trees come up for the animal to graze and browse on. Early burning gives woody vegetation an advantage over the grasses, while late burning encourages the growth of grass by discouraging the woody vegetation (Geerling, 1973). Fire, then, has considerable potential for manipulating the habitat according to requirements.

It is reasonable to suppose that grazing animals will benefit from burning whilst browsers will do better with some tree and shrub cover in their habitat. Generally, prolonged use of fire in the savannah has resulted in the development of special fire-tolerant communities of plants and animals which are dependent on periodic burning for their existence (Glover, 1968).

It can be concluded that through skilful protection or controlled burning it is possible to reduce the destructive effects of heavy animal populations on forest remnants and good grazing can be restored by encouraging fires late in the dry season. By rational burning it is possible to ensure abundance of tender grass and to attract game to popular viewing points.

THE ROLE OF SALT LICKS IN WILDLIFE CONSERVATION

Vertebrates have complex nutritional requirements in the form of chemical elements. As such, like water and good supply of food, the salt lick constitutes one of the requirements expected in an ecological unit. Not all animals regularly come to salt licks, but observations have revealed that almost all the big game pay some visits, e.g. elephants, big antelopes, baboons (Henshaw and Ayeni, 1971).

It is not yet certain which chemical element is most sought at the licks. It has been suggested, however, that trace elements may be the critical constituents and generally phosphorous and sodium are believed to be the two principal trace elements causing animals to use licks (Cowan et al., 1949).

Henshaw and Ayeni (1971) observed that no single element from the mineral licks sampled in Yankari, can be isolated to demonstrate a specific preference on the part of the species commonly utilizing the licks. They also observed that levels of minerals such as nitrogen, potassium and phosphorus are undoubtedly increased at the licks due to urination and defecation. Due to accelerated growth rates, calcium and, possibly, phosphorus may be particularly desired during the rainy season. In this respect it is pertinent that use of salt licks by most species increases during the rains and that western hartebeest, which generally feed in the drier, less mineral rich areas of the reserve, particularly favour the licks.

Degradation of the habitat through over-grazing, over-browsing and soil compaction results from heavy lick use. There is also an increase of disease and predation often leading to high mortalities. This is illustrated by the frequency of carcasses near licks.

Knowledge of lick positions in a conservation area, greatly helps in determining animals' movements and distribution. In Yankari Game Reserve hides are installed near some of the prominant licks. Visitors and research workers wanting to take pictures of wildlife at close range or to make behavioural observations find these hides very useful. Viewing tracks are generally constructed to pass the main licks and impressive animal sightings during viewing trips are made on such tracks. Licks are often visited by poachers. So anti-poaching patrols are often aimed at these areas.

THE ROLE OF WATER REGIME IN WILDLIFE CONSERVATION

Wildlife species vary greatly in their need for drinking-water. Some are adapted to live with very little of it, like the roan antelope in the Yankari Game Reserve. Some animals do not even drink at all, like oryx. However, almost all animals and birds benefit directly or indirectly from the presence of water in a habitat. Big game graze and browse on fresh vegetation and of course its growth is encouraged by the proximity to water.

In the Ngaserai Control Area in Tanzania, wildlife and livestock are allowed to co-exist. The area is dry and in the rain-shadow of Mount Kilimanjaro. Sources of water are very scarce, especially during the dry season, and watering points were developed for the Masai livestock, which attracted a large wildlife population from the neighbouring areas. As a result the concentration of both the wildlife and the livestock became very high. The nearby College of Wildlife was able to make use of the area for demonstrations and specimen collection for their students, and a hunting club was given a concession in the area.

In Yankari Game Reserve there is only one major source of water — the Gaji river, which runs across the reserve. During the rainy season, the animals obtain good grazing and browsing and drinking-water all over the reserve. During the dry season, most of the watering points become dry except the Gaji river and its valleys. This is the permanent watering point for the animals. As a result there is a seasonal high concentration of animals along the Gaji river valleys. This is very advantageous because visitors can easily view the animals in large herds. However as soon as it rains, the animals disperse and become more difficult to see.

In the developing Lame/Burra Game Reserve, Nigeria, there are many rivers and streams that retain water throughout the year.

In the Ngaserai Controlled Area, the influx of the wildlife into the area as a result of the water sources, has brought about direct competition between wildlife and livestock. The competition is mostly between the larger wildlife species such as wildebeest and zebra. The smaller species such as Thomson's gazelle are involved little in competition because they favour the scanty grasses that are dominant in the area. In addition the influx is causing damage through overgrazing and 'local' overstocking. The animals gather near the water sources in large number and thus encourage soil erosion while animal mortality is increased through fighting, predation and rapid spread of disease.

At Yankari Game Reserve, seasonally high concentrations of animals along the Gaji River result in damage to the riverine vegetation and also to soil erosion. The elephants usually congregate in herds of 150–200 grazing, browsing, pushing trees down and compacting the fragile soil by trampling. Here, again, mortality amongst the animals is high due to predation, fighting, disease and poachers. Opinions on the situation at Ngaserai vary from views that the whole area is overstocked, to views that it is only 'locally' overstocked (Lasan, 1971). An increase in the number of watering points will only result in increasing density of animals in the area. This does not solve the problem. Instead, it intensifies it. The solution is to keep the number of the animals within the carrying capacity of the area. The fact that damage is intensive in the form of destruction of vegetation and increased incidence of soil erosion, indicates that the carrying capacity has been exceeded. According to Dassman (1964) once water supplies have been developed to a maximum extent it is essential that the animal numbers be limited to the carrying capacity of the area within travelling distance from the watering places.

The situation in Yankari Game Reserve is the opposite. Here, spreading the animals to give relief to the Gaji valley seems the logical solution. This can be done by developing the existing waterholes spread through the reserve, to retain water all the year round. This will also call for intensification of anti-poaching patrols since the animals will be found by the poachers throughout the area all the year round. Such action will considerably reduce the number of animals in the main viewing areas, but there will still be enough to view because the lush vegetation and abundant water supply will still attract animals. The animals themselves will be relieved of their social stress and thus their rate of reproduction will probably increase. The Lame/Burra Reserve, on the other hand, presents a different situation completely since there is water all year, and in most parts of the area the wildlife population if evenly distributed. Thus, animals will rarely be encountered in such high concentrations as in Yankari. Increases in population will appear meagre and an extensive census method will be needed to demonstrate them.

REFERENCES

BARTLETT, H. H. (1957) *Fire in relation to primitive agriculture and grazing in the tropics*, Vol. 2. (Ann. Arbor, Michigan).

COWAN, I. MCTAGGART, and BRINK, V. C. (1949) Natural game licks in the Rocky Mountains national parks of Canada. *Journal of Mammalogy*, Vol. 30, pp. 387–99.

DASSMAN, R. F. (1964) *Wildlife biology.* (John Wiley, New York).

DOUGALL, H. W., and GLOVER, P. E. (1964) On the chemical Composition of *Themedra triandra* and *Cynodon dactylon*. *East African Wildlife Journal*, Vol. 2, pp. 67–70.

GEERLING, C. (1973) The vegetation of Yankari Game Reserve: its utilization and condition. (Department of Forestry, University of Ibadan, Bulletin 3.)

GLOVER, P. E. (1963) The role of fire and other influences on the savannah habitat, with suggestions for further research. *East African Wildlife Journal*, Vol. 6, pp. 131–7.

HENSHAW, J. and AYENI, J. S. O. (1971) Some aspects of big/game utilization of mineral licks in Yankari Game Reserve, Nigeria. *East African Wildlife Journal*, Vol. 9, pp. 73–82.

LASAN, A. H. (1971) *Formulation of a management plan for the Hongido controlled area, Ngaserai.* Unpublished MS, (College of African Wildlife Management, Mweka).

MESS, M. G. (1958) The influence of veld burning on the water, nitrogen and ash content of grasses. *East African Journal of Science*, Vol. 54, pp. 83–6.

12

THE NUTRITIVE VALUE OF BROWSE AND ITS IMPORTANCE IN TRADITIONAL PASTORALISM

W. L. Brinckman and P. N. de Leeuw

National Animal Production Research Institute,
Ahmadu Bello University, Zaria, Nigeria

INTRODUCTION

Some 60 million ha of uncultivated land in the northern part of Nigeria can potentially be used for grazing and livestock production (de Leeuw, 1974). These lands carry a savannah vegetation, consisting of an upper stratum of trees and shrubs of variable density and a herbaceous cover mainly of grasses.

If all savannah were available for grazing and managed efficiently, there would be sufficient forage to feed the present national herd of ten million cattle and over ten million sheep and goats, and also to allow for rapid expansion. However, uneven distribution of water supplies, tsetse infestation, disease risks and difficulties of access limit optimum utilization and have also resulted in serious overstocking of the Sudan zone rangelands and under-utilization of areas further south. Some of these constraints will, no doubt, be overcome with the continued development of the livestock industry.

Whatever is, or will be, done to improve rangeland utilization and increase fodder supplies, there is no easy remedy to the extremely poor quality of most grass herbage in the dry season, except through eliminating the tree cover with heavy equipment and converting the savannah into sown legume/grass pastures. As in all ecosystems that are subject to monsoonal climates, fodder quality is satisfactory during the 3–5 months growing season but digestibility and nutrient content decline rapidly when grasses mature and become dry and fibrous, so that for a long period stock must subsist on a below-maintenance diet. This prolonged dietary deficiency in nutrients (particularly protein and phosphorus) is held largely responsible for retarded growth, late maturity, poor milk yields and low reproductive rates in Nigerian zebu cattle and appears also to be the main factor limiting wild herbivore populations in East African ecosystems (Sinclair, 1975).

Although during the dry season, fibrous standing grass is the major fodder source, herd owners try to vary the diet of their stock by making available more

101

nutritious fodder (browse, hay from leguminous crops, green regrowth in low-lying or burned areas), as they fully realize that such supplementation benefits their stock (van Raay and de Leeuw, 1970, 1974). Among these better fodder browse ranks first: leaves, flowers and fruits of many trees and shrubs are readily consumed. This paper reviews the nutritional value of browse and tries to evaluate its contribution to the nutrition of livestock within the context of traditional semi-nomadic and sedentary systems of livestock husbandry.

MINERAL AND PROTEIN CONTENT

The mineral requirements of cattle are far from accurately known, and feeding standards often include only calcium and phosphorus. According to the *Nutrient requirements of beef cattle* (A.R.C., 1965), rations should contain at least 0.18 per cent of the dry matter as calcium and phosphorus. The ratio between calcium and phosphorus should ideally fall between 1:2, but ratios of 7:1 in feeds have been reported as satisfactory. Low phosphorus levels in pasture herbage and other roughages are widespread particularly in semi-arid regions, and are commonly associated with phosphorus-deficient soils. Phosphorus content of plants generally decreases with maturity and deficiencies often occur in mature dried forage as is the case in northern Nigeria (see Table 1). Symptoms of phosphorus deficiency are decreasing appetite and reduced growth.

Magnesium requirements of cattle are reported to be between 12 and 30 mg/kg body weight (A.R.C., 1965) which is comparable to 0.05–0.12 per cent in the feed. This level is reached in most fodders. Potassium requirements of cattle have also not been critically measured, but the optimum levels for growing cattle have been reported (A.R.C. 1965) to be 0.6–0.8 per cent of the dry matter of the feed. Deficiency results in non-specific symptoms such as slow growth, reduced feed consumption and efficiency, stiffness and emaciation.

Table 1 gives average contents of four macro-elements in grasses and browse plants, together with the estimated needs of cattle. For calcium, magnesium and potassium, average values are adequate. All the authors found averages higher than needed. Browse showed higher levels than grasses, but there were no indications of toxic levels. Calcium/phosphorus ratios are rather high, especially for browse. Phosphorus values are low, five out of seven values in grass and two out of five values for browse species are below the minimum. Phosphorus contents which are highly variable, are higher in browse than in grasses and higher in wet-season grass than in dry-season grass (see Table 2). Data from Kenya indicate that phosphorus deficiency in East Africa is less striking than in West Africa (Dougall, Drysdale and Glover, 1964).

The crude protein (CP) contents of grasses and browse are compared in Table 3. It is clear that browse plants are superior to grasses. Especially during the dry season, nearly all the grasses have a CP content below 5 per cent, whereas 5–6 per cent is assumed to be the minimum protein requirement for

TABLE 1

Calcium, phosphorus, magnesium and potassium content of grass and browse in West Africa

Source	Ca	P	Mg	K	Ca/P	‖
Grasses						
Kapu (1975)†	0.34	0.22	0.23	2.56	1.55	9
Kapu (1975)§	0.22	0.03	0.11	0.37	7.33	5
Boudet & Ellenberger (1971)	0.44	0.14	0.32	1.35	3.12	34
Bartha (1970)‡	0.31	0.22	0.22	2.12	1.41	36
Rippstein & Peyre de F. (1972)	0.43	0.11	0.24	1.43	3.91	66
Peyre de Fabreques (1967)	0.40	0.15	0.18	1.43	2.75	20
Audru (1966)	0.28	0.11	0.19	0.94	2.48	32
Browse						
Kapu (1975)§	0.70	0.21	0.28	1.66	3.33	10
I.A.B. (1947)	1.62	0.26	—	1.63	6.23	47
Bartha (1970)	1.23	0.15	0.40	1.65	8.04	20
Rippstein & Peyre de F. (1972)	2.28	0.22	0.57	2.32	10.22	35
Audru (1966)	1.53	0.15	0.60	1.45	10.45	27
Average grasses	0.35	0.14	0.21	1.46	3.22	202
Average browse	1.47	0.20	0.46	1.79	7.65	139
Estimated needs (ARC, 1965)	0.28	0.19	0.07	0.25	1.47	

†Wet season.
‡Late wet season.
§Late dry season all other data for the whole year
‖Number of samples.

TABLE 2

Frequency distribution of phosphorus content in grasses and browse in West and East Africa

	Grasses			Browse		
P content % D.M.	Sahel/ Sudan dry season	Sudan wet season	Sahel/ Sudan	Northern Guinea	East Africa
0 –0.05	10	—	3	2	—
0.06–0.10	29	36	15	22	6
0.11–0.15	25	43	34	27	19
0.16–0.20	17	13	23	17	30
0.21–0.25	4	8	14	15	22
Over 0.25	14	—	11	17	23
No. of samples	138	53	92	40	95

(After Dougall *et al.*, 1964.)

TABLE 3

Frequency distribution (%) of crude protein content (% DM)† in grasses and browse

	Grasses			Browse		
CP content	Sahel dry season	Guinea dry season	Sudan wet season	Sahel	Guinea Year-round	East Africa
0– 2.5	27	25	15	—	—	—
2.0– 5.0	50	71	41	2	—	—
5.1– 7.5	16	3	21	4	10	6
7.6–10.0	6	1	21	8	24	9
10.1–12.5	1	—	2	23	27	26
12.6–15.0	—	—	—	15	17	18
15.1–17.5	—	—	—	16	15	16
17.6–20.0	—	—	—	11	2	9
Over 20.0	—	—	—	21	5	16
No. of samples	176	273	53	107	41	97

†DM = Dry matter (%).

maintaining body weight (Crampton and Harris, 1969). In Shika 85 per cent of the grasses sampled contain less than 4 per cent CP and 50 per cent less than 3 per cent. Average CP of browse species is well above maintenance requirements.

In Table 4 the CP content of different parts of the grass vegetation is illustrated. If animals select the tops of the grass (above 30 cm) then they will be able to cover their protein needs during 5–6 months, although average values for the whole plants are rather low.

THE CONTRIBUTION OF BROWSE TO PROTEIN NUTRITION

Since it has been shown that the average browse is superior to grasses as regards CP content, its role in protein nutrition for livestock needs assessment, the more as protein deficiency is believed to be the major cause for the low productivity of traditionally managed herds (Zemmelink, 1973).

The effect of the dry-season protein deficiency in growing stock has been amply demonstrated in grazing trials in which stock grazed the roughage available in northern Guinea shrub savannah and were given little or no supplemental protein. As this roughage contained no digestible, protein, stock lost weight to the tune of 15 per cent of the body weight they had attained at the end of the rainy season (Zemmelink, 1973; de Leeuw, 1971; Hagger, de Leeuw and Agishi, 1971).

To what extent dietary protein deficiency affected traditionally managed herds, was studied by van Raay and de Leeuw (1970, 1974). They conducted a

TABLE 4

Crude protein content (% DM) of grass herbage components in burned and protected shrub savannah in the Northern Guinea Zone (Shika)

Plant component		Sampling Period						
		May 26–28	Jun. 26–28	Jul. 23–26	Aug. 18–23	Sep. 18–23	Oct. 21–24	Dec. 3–8
Green herbage 0–30 cm	burned	13.5	6.6	5.0	3.6	3.0	3.3	2.9
	protected	8.9	6.2	5.5	3.8	3.3	3.9	3.2
over 30 cm	burned	—	—	—	5.7	5.2	5.1	3.7
	protected	—	—	—	5.7	5.0	5.4	4.1
Dry and dead herbage	burned	3.2	—	2.7	2.4	2.1	1.9	2.1
	protected	1.9	1.9	1.9	1.9	2.0	2.8	2.0
Flowering culms	burned	—	3.8	2.4	2.3	3.1	2.6	1.7
	protected	1.7	2.7	2.1	2.0	3.2	3.0	2.0
Total herbage	burned	12.4	6.0	4.7	4.1	4.0	3.3	2.2
	protected	2.6	4.0	4.0	4.1	3.7	3.5	2.2

(After De Leeuw, unpublished.)

year-round exploratory analysis of the grazing practices of semi-nomadic Fulani in Katsina and sedentary herd owners in the Zaria area, Northern Nigeria. This involved day-long time studies of herd activities (duration and intensity of grazing, walking speed and distance, frequency of watering etc.) while the major components of the grazing diet (grass, herbs, browse and different crop residues) were recorded and time spent on each source measured, throughout each sample day.

From this time analysis it was attempted to estimate the monthly average digestible crude protein (DCP) intake per day for the two management systems under study. The following procedure was followed:

(1) Diet composition was estimated from the total time animals spend on each fodder source, assuming a linear relationship between grazing time and intake.
(2) Daily feed intake was assumed to be constant and taken as 7 kg of dry matter for an adult beast with a body weight of 250–300 kg.
(3) The average DCP content for all fodder components was calculated from CP data collected during the survey. The formula DCP $\% = 0.957$ CP $\% - 3.75$ (Milford and Minson, 1965) was used for this conversion. The average CP and DCP contents of the browse were 14.4 and 10.0 per cent respectively.
(4) The maintenance requirements of DCP for an adult zebu was taken as 150 g/day, while an allowance for walking was added at a rate of 4 g/day/km (Boudet and Riviere, 1967).

The average monthly DCP intakes in relation to the estimated requirements for maintenance is given in Table 5. These confirm the expected protein deficit during six months in the semi-nomadic and during nine months for the sedentary herds, although the latter are less exposed to the extreme variations in their diets. Browse as a protein source is much more important in Katsina than in Zaria, because the semi-nomadic Fulani move mainly in well-wooded grazing reserves with many browse species, while the sedentary herds must rely mainly on fallow-land and flood-plains because their grazing orbit is centred in a densely populated area.

Although browse in the diet falls far short of making up the dry-season protein deficit, its relative contribution is substantial particularly in Katsina during March and April.

DISCUSSION

The comparisons between the nutrient content of grasses and browse showed that, on average, browse contained much more crude protein than grasses — so that given the highly selective grazing habits of zebu cattle, it may be postulated that they select browse plants containing more than 10 per cent crude protein.

The impact of dry-season protein supplementation on cattle performance is well documented. Zemmelink (1973), for instance, reported that 100–130 g DCP in concentrates per day would maintain weight in a 250 kg steer, which would loose 300–400 g daily, if not supplemented. This supplement provided only 10 per cent of the maintenance energy requirement, the remainder being supplied by the roughage. In other words, protein supplementation is the key to efficient utilization of low quality dry-season fodder resources, which at least on a regional scale, are in plentiful supply. Consequently, browse should be regarded as a protein supplement particularly during the period of greatest stress. In April and May browse provided up to 100 g of DCP which is comparable to 750 g of cotton-seed or 250 g of ground-nut cake.

Since more than half of the browse samples have less than the recommended phosphorus content (see Tables 1 and 2), selective grazing of browse plants is less likely to make up for deficiencies. Thus, year-round supplementation with mineral mixtures rich in phosphorus have received attention (Umoh and Koch, 1971).

Rates of browsing vary considerably and depend on the tree and shrub composition of the vegetation, the relative availability of other sources of fodder and the time of year. Cattle, and to a lesser extent sheep, prefer grasses to other forage. When grass is abundant, it usually accounts for more than 80 per cent of the total intake. Where grass is scarce, browse is known to become the major source of fodder (Connor et al., 1963; Rees, 1974).

In summary, it can be stated that browse will continue to play an important role in the nutrition of traditionally managed livestock, since during several months in the dry season, it may be the only source of dietary protein. Although

TABLE 5

The contribution of browse to the digestible crude protein (DCP) intake of cattle in two pastoral management systems

Month	Dec.	Jan.	Feb.	Mar.	Apr.	May	Jun.	Jul.	Aug.	Sep.	Oct.	Nov.
Kasina: Semi-nomadic herds												
Browsing % total grazing	—	1	2	15	16	15	20	13	2	6	10	11
DCP intake from browse g/day		7	14	105	112	105	140	91	14	42	70	77
% of total		6	18	81	100	25	47	34	6	22	33	45
Total DCP intake: % of estimated maintenance requirements	50	57	37	50	46	238	121	239	103	99	122	89
Zaria: Sedentary herds												
Browsing, % total grazing	—	2	3	5	6	6	2	4	2	5	10	2
DCP intake from browse g/day	—	14	21	35	42	42	14	28	14	35	70	14
% of total	—	18	27	29	31	14	3	4	10	30	48	17
Total DCP intake: % of estimated maintenance requirements	78	39	88	60	67	169	288	126	81	60	75	43

(Derived from van Raay and de Leeuw 1974.)

other high-protein feeds are available in Nigeria (cotton-seed, cotton-seed cake and ground-nut cake), which could replace browse, these are expensive and not easily distributed to where they are needed. In addition, the inborn conservatism of most pastoralists seems to preclude the large-scale adoption of dry-season supplementation with purchased concentrates. Thus, as long as the majority of livestock are in the hands of the pastoral tribes, rangeland programmes that include the total destruction of the tree cover should be viewed critically, unless it is fully proven that the forage cover, replacing woodland savannah, is superior in nutritive value, in particular during the later part of the dry season.

REFERENCES

AGRICULTURAL RESEARCH COUNCIL (1965) *Nutrient requirements of Livestock.* Part 2: Ruminants.

AUDRU, (1966) Etude des Qaturages et des Problemes Pastoraux dans le Delta du Senegal. *Etude Agrostologique No. 15.* (I.E.M.V.T. Maisons-Alfort, France).

BARTHA, R. (1970) Fodder plants in the Sahel zone Africa. CFO-Institut, Munich) p. 306.

BOUDET, G. and ELLENBERGER, J. F. (1971) Etude agrostologique du berceau de la race N'dama dans le Cercle de Yanfolila, Mali. *Etude Agrostologique No. 30.* (I.E.M.V.T. Maisons-Alfort, France).

BOUDET, G. and RIVIERE, R. (1967) Emploi d'analyses fourrageres pour l'appreciation de paturages tropicause. (I.E.M.V.T. Maisons-Alfort France).

CONNER, J. N., BOHMAN, U. R., LESPERANCE, A. L. and KINSINGER, F. E. (1963) Nutritive evaluation of summer forage with cattle. *Journal of Animal Science*, Vol. 22, pp. 961–9.

CRAMPTON, E. W. and HARRIS, L. E. (1969) *Applied Animal Nutrition.* (Freeman & Co., San Francisco).

DOUGALL, H. W., DRYSDALE, V. M. and GLOVER, P. E. (1964) The chemical composition of Kenya browse and pasture herbage. *East African Wildlife Journal*, Vol. 2, pp. 86–121.

HAGGER, R. J., DE LEEUW, P. N. and AGISHI, E. (1971) The production of *Stylosanthes gracilis* at Shika, Nigeria, II. In savannah grassland. *Journal of Agricultural Sciences*, Vol. 77, pp. 437–44.

IMPERIAL AGRICULTURAL BUREAU (1947) *The use and misuse of shrubs and trees as fodder.*

KAPU, M. M. (1975) The natural forages of Northern Nigeria. 1. Nitrogen and mineral composition of grasses and browse from the northern Guinea savannah and standing hays from the different savannah zones. Paper presented at the third conference of the Nigerian Society of Animal Production, Enugu 1975.

LEEUW, P. N., DE (1971) The prospects of livestock production in the Northern Guinea Zone. *Samaru Agricultural Newsletter*, Vol. 13, pp. 124–33.

LEEUW, P. N. DE (1974) Livestock development and drought in northern states of Nigeria. *Nigerian Journal of Animal Production*, Vol. 1, pp. 61–73.

MILFORD, R. and MINSON, D. J. (1965) Intake of tropical pasture species. *Proceedings of the 9th International Grassland Congress.* Sao Paulo, pp. 815–22.

NATIONAL ACADEMY OF SCIENCES (1970) *Nutrient Requirements of Beef Cattle.* (Washington, D.C.).

RAAY, J. G. T., VAN and DE LEEUW, P. N. (1970). The importance of crop residues as fodder: A resource analysis in Katsina Province, Nigeria. Vol. 40, pp. 137–47.

RAAY, J. G. T., VAN and DE LEEUW, P. N. (1974) Fodder resources and grazing management in a savannah environment: an ecosystem approach. (Institute of Social Sciences, The Hague) p. 29.

RIPPSTEIN, G. and PEYRE DE FABREQUES, B. (1972) Modernisation de la zone pastorale du Niger. *Etude Agrostologique No. 33.* (I.E.M.V.T. Maisons-Alfort, France).

REES, W. A. (1974) Preliminary studies into bush utilization by cattle in Zambia. *Journal of Applied Ecology*, Vol. 11, pp. 207–14.

PEYRE DE FABREQUES, B. (1967) Etude agrostologique des paturages de la zone nomade de Zinder, Republique du Niger. (UNDP/IEMVPT, Maison Alfort, France).

SINCLAIR, A. R. E. (1975) The resource limitation of trophic levels in tropical grassland ecosystems. *Journal of Animal Ecology*, Vol. 44, pp. 106–9.

UMOH, J. E. and KOCH, B. A. (1971) Preliminary salt and mineral feeding studies at ARS, Shika. *Samaru Agricultural Newsletter*, Vol. 13, pp. 106–9.

ZEMMELINK, G. (1973) Utilization of poor quality roughages in the northern Guinea savannah zone. *Animal Production in the Tropics* (Heinemann) pp. 167–76.

13

SPECIES PREFERENCES OF DOMESTIC RUMINANTS GRAZING NIGERIAN SAVANNAH

P. N. de Leeuw

National Animal Production Research Institute,
Ahmadu Bello University, Zaria, Nigeria

INTRODUCTION

In recent studies on the grazing management of pastoral herds it was shown that traditional systems of husbandry have evolved in which the provision of a varied fodder diet to stock is a predominant aim. Herdsmen living in close touch with their savannah environment take continuous advantage of the seasonal availability of all fodder sources and adapt their management and adjust their movement to this ever-changing supply situation. Over centuries they have accumulated detailed knowledge of the fodder potential in their regions of orbit and know the grazing habits and fodder preferences of their stock (van Raay and de Leeuw, 1970, 1974).

Species preferences of domestic stock grazing savannah lands in West Africa are particularly well-documented for the Sahel and Sudan zones, where most herds have been concentrated due to the tsetse risk further south. These reports indicate the wide range of grasses, herbs and woody plants that are consumed. There is much less information on the palatability of savannah plants in the Guinea zone, although this zone is attracting increasing numbers of herds concomitant with the gradual reduction of the tsetse risk. These southward movements have intensified since the 1972–73 drought and there is much evidence that many pastoralists have permanently settled in these hitherto ungrazed savannahs.

Most studies describing the species composition and the productivity of savannah point out the higher nutritive value of browse and emphasize its importance as a source of protein and minerals, particularly in the dry season when grasses are poorly digestible and are poor in nutrients. However, to quantify the contribution different species or even groups of species make to the grazing diet is extremely difficult. While Rose-Innes and Mabey (1964) found that cattle in Ghana could ingest rations with very high browse content and Rees (1974) in Zambia claimed that steers could sustain high gains in

110

pollarded savannah woodland, Wilson (1969) in a review on the importance of browse stated that 'it has been shown that browse has an important contribution to make to the nutrition of domestic and most game animals'.

This paper reports on a series of studies in which browsing time was estimated and on species preferences of cattle and sheep grazing the Guinea and Sudan savannahs under different systems of management.

BROWSING TIME

Cattle confined to paddocks

During 1966–67 a series of animal behaviour studies were carried out, in Shika near Zaria in northern Nigeria using the methodology outlined by Haggar (1968). Activities such as grazing, walking, ruminating, standing and idling were recorded at intervals of 2.5 minutes over 24 hour periods. Time spent on grazing herbaceous forage and on browsing trees and shrubs was recorded separately. These records were taken during four distinct periods: early rainy season (2–21 June), rainy season (12–30 August), mid dry season (4–14 February) and late dry season (7–18 April). The schedule of observations is presented in Table 1.

TABLE 1

Schedule of observations on animal behaviour

Period	Jun.	Aug.	Feb.	Apr.
Number of paddocks	4	4	4	4
Number of animals	12	8	8	8
Days of observation/head	2–3	4–5	2	3
Total number observations/period	34	34	16	24

Four groups of 3 zebu bulls of about 250 kg weight, each grazed 8 ha shrub savannah subdivided into 4 paddocks. In June all the bulls grazed and in the other periods 2 out of 3 bulls were observed. During the study shrubs were counted in 80 sample plots (400 m² in size). The mean frequency per species is given in Table 2, which shows that *Isoberlinia doka* and *Dichrostachys cinerea* account for over 40 per cent of the total tree stand.

Browsing time varied considerably between periods, being negligible in February and up to 1.5 hours in April (see Table 3). In February most shrubs are leafless in contrast to April when all have new leaves. In Table 2 the species have been listed in three palatability classes. It can be seen that over half the arboreal cover consisted of shrubs that are not consumed by stock. Of the

TABLE 2

Relative density (%) and palatability of trees and shrubs in upland shrub savannah at Shika

Readily consumed species		Occasionally consumed species	
Adenodolichos paniculatus	0.7	Anogeissus leiocarpus	0.6
Bridelia ferruginea	0.7	Burkea africana	1.5
Butyrospernum paradoxum	1.6	Combretum ghasalense	1.2
Combretum aculeatum	0.2	Cussonia barteri	0.5
Dichrostachys cinerea	13.8	Daniellia oliveri	0.4
Feretia apodanthera	0.2	Parinari curatellifolia	0.9
Grewia mollis	1.6	Pavetta crassipes	1.2
Parkia clappertoniana	0.2	Psorospernum febrifugum	1.7
Stereospernum kunthianum	0.6	Pterocarpus erinaceus	2.4
Piliostigma thoningii	2.4	Securidaca longepedunculata	0.7
Vitex doniana	0.3	Strychnos spinosa	4.0
		Terminalia avicennioides	9.3
		T. laxiflora	2.2
		Swartzia madagascarensis	0.8
Total	22.3	Total	27.4

Avoided species			
Afrormosia laxiflora	2.3	Lippia chevalieri	2.3
Albizzia chevalieri	0.8	Monotes kerstingii	0.9
Annona senegalensis	5.2	Ochna afzelii	0.7
Cassia singueana	0.6	Steganoteania araliacea	1.2
Detarium microcarpum	1.6	Terminalia glaucescens	0.5
Entada africana	0.8	Trichilia emetica	0.8
Gardenia cf. erubescens	0.4	Ximenia americana	0.9
Isoberlinia doka	28.6	Other species	1.9
		Total	50.3

TABLE 3

The seasonal variation in grazing and browsing time of confined cattle in Shika

Period	Grazing and browsing hrs±S.E.	Browsing only	
		min.+S.E.	%±S.E.†
Feb.	8.24±0.33	18±4	3.6±0.8
Apr.	9.55±0.35	70±7	12.2±1.3
Jun.	9.91±0.20	89±6	15.0±1.1
Aug.	8.57±0.28	44±2	8.6±0.6

†of total grazing and browsing time.

shrubs browsed *Dichrostachys* and *Termininalis avicennoides* are the most common, while the other species in the same class occur only in limited numbers.

Herded cattle

The recording of browsing time was included in the analysis of the grazing management of traditionally managed cattle in the Katsina and Kano states of Nigeria. Zaria area (van Raay and de Leeuw, 1974). Table 4 shows that browsing is more important in Katsina than in Zaria, probably because the Katsina herds move around in open savannah woodland which harbour many palatable browse species (see Table 6) in contrast to the Zaria herds, which were largely restricted to fallowland and low *Isoberlinia Terminalia* shrub savannah with less shrubs likely to be browsed. As in Shika, the peak in browsing coincided with the flush of new leaves at the end of the dry season. In both grazing areas there was little browsing between December and February, since herds during that period were mainly in farmland to graze the residues of sorghum and other crops after harvest (van Raay and de Leeuw, 1970).

PALATABILITY OF GUINEA ZONE GRASSES

The palatability of the grass species occurring in the upland shrub savannah of the Agricultural Research Station, Shika were studied during the 1967 and

TABLE 4

The seasonal variation of grazing and browsing time in two herds

Herd ecological zone	Nomadic Sudan			Sedentary Guinea		
	Grazing and browsing only			Grazing and browsing only		
Months	browsing hrs.	min.	%	browsing hrs.	min.	%
Dec.	7.2	—	—	7.1	—	—
Jan.	7.1	4	1	7.3	9	2
Feb.	6.5	8	2	7.0	13	3
Mar.	7.4	67	15	7.0	21	5
Apr.	8.2	79	16	7.4	27	6
May	9.9	89	15	7.2	26	6
Jun.	8.3	100	20	6.8	8	2
Jul.	7.6	59	13	6.5	16	4
Aug.	7.0	8	2	6.3	8	2
Sep.	7.0	25	6	6.2	19	5
Oct.	7.2	43	10	6.4	38	10
Nov.	7.4	49	11	6.8	8	2

(Adapted from Van Raay and de Leeuw 1974.)

1968 rainy seasons. Over this period nine paddocks were sampled once or twice, giving a total of 15 palatability recordings; seven were carried out in August and four each in September and October. On each date 50 quadrats (0.35 m² in size) were laid out at regular intervals along two parallel lines. The species composition was measured by ranking the three important species and according to these scores of 3, 2 and 1, and by noting the presence of additional species. Subsequently it was noted which in each quadrat were grazed.

The paddocks (2.4–4.0 ha in size) used for sampling were rotationally grazed at an overall stocking rate of 0.75 head per ha by two groups of 12 zebu and zebu–Friesian crosses weighing 200–250 kg. They were moved every 14 days and the date of sampling was close to the end of the grazing period.

The vegetation records enabled the computation of relative dominance and frequency for each species and its relative palatibility (the ratio between frequency of grazing and frequency of occurrence). Overall means and means for August and September–October have been tabulated in Table 5. The values expressing the species composition correspond closely to the rate grazed. This indicates that cattle when confined to a fenced land do not do much preferential grazing and consume the majority of species present. *Andropogon* spp and *Panicum phragmitoides* are better liked than *Setaria* and *Hyparthelia dissoluta*. *Urelytrum* and *Elyonurus pobeguinii* are relatively unpalatable. Most species of low occurrence are eaten, which seems to infer that cattle like a varied diet. Although differences in preferences between periods are small, they become greater when the grass cover is reaching maturity.

THE PALATABILITY OF TREES AND SHRUBS

In Shika

Species preferences of herded cattle grazing upland savannah were recorded from June 1965 to January 1966. Groups of mature cows, young bulls or heifers, were followed by two to four experienced vegetation recorders, who were familiar with the savannah flora. As the animals were used to herded management, they could be easily approached and plants eaten observed and identified. Eleven recordings were made throughout this period.

In order to arrive at comparative palatability ratings for individual species, each time that a plant was eaten or nibbled was considered a 'hit'. Percentage score per species relative to the total number of hits per sampling could therefore be calculated. Depending on the duration of the observation period (3–5 hours) the number of heads in the herd (30–50 heads) and the amount of browse available, hits per sample ranged from 60 to 500 totalling 2650 hits for the entire sampling period. Two classes of plants were established: plants that were readily consumed throughout the sampling period and those eaten only occasionally (see Tables 2 and 6).

TABLE 5

Relative palatability of grasses in relation to species composition of shrub savannah

Species	Aug.			Sep.–Oct.			Mean			
	D%¹	F%²	g%³	D%¹	F%²	g%³	D%¹	F%²	g%³	g%/F%⁴
Andropogon gayanus	6.6	6.1	9.2	8.0	7.4	12.1	7.4	6.9	10.9	1.6
A. schirensis	2.9	3.6	6.0	5.3	5.1	8.6	4.3	4.5	7.6	1.7
A. ascinodis	5.1	5.6	8.1	6.4	6.7	8.2	5.0	6.3	8.1	1.3
Panicum phragmitoides	11.7	11.3	11.5	11.6	10.1	14.5	11.7	10.7	13.3	1.2
Setaria anceps	13.0	12.3	14.8	12.4	12.9	13.7	12.8	12.8	14.2	1.1
Hyparrhelia dissoluta	17.1	12.0	15.6	8.4	7.6	7.7	12.0	9.5	10.8	1.3
Urelytrum muricatum	12.8	9.8	9.3	14.2	10.6	8.6	13.7	10.2	8.9	0.9
Elyonurus pobeguinii	12.7	12.8	3.8	15.3	14.0	3.7	14.2	13.5	3.8	0.3
Ctenium newtonii	5.0	6.5	4.7	3.2	4.8	2.6	3.5	5.5	3.5	0.6
Monocymbium ceresiiforme	1.5	2.5	2.5	2.6	3.8	4.3	2.2	3.3	3.6	1.1
Brachiaria jubata	1.1	1.5	2.0	1.5	2.1	3.0	1.3	1.9	2.6	1.4
Sporobolus pyramidalis	2.9	4.0	4.5	1.5	2.4	1.2	2.1	3.0	2.6	0.9
Hyparrhenia suplumosa	0.8	1.2	1.8	2.5	2.9	4.0	1.9	2.2	3.1	1.4
Cymbopogon giganteus	1.6	2.1	0.6	1.7	2.0	1.2	1.6	2.1	1.0	0.4
H. bagirmica	0.7	1.3	1.1	0.5	0.8	0.8	0.6	1.0	0.9	—
H. involucrata	0	3	0	3	0.6	0.5	0.7	0.3	0.4	0.5
Pennisetum pedicellatum	0.4	0.5	0.5	1.6	1.7	2.4	1.1	1.3	1.6	—
P. polystachion	0.3	0.6	0.7	0.4	0.7	0.6	0.6	0.4	0.7	—
Paspalum sp.	1.6	3.0	2.0	0.2	0.4	0.4	0.8	1.4	1.0	—
Andropogon pseudapricus	—	—	—	0.6	0.8	0.7	0.3	0.5	0.4	—
Sporobolus festivus	1.0	1.6	—	0.4	0.7	—	0.6	0.6	—	—
Elyonurus hirtifolius	0.4	0.3	—	0.4	—	0.4	0.3	0.4	—	—
Other species	0.5	1.1	0.7	0.8	1.4	1.4	0.8	1.4	1.0	—

1. D% Relative dominance.
2. F% Relative freqency of occurrence.
3. g% Relative frequency of grazing.
4. Relative palatability.

TABLE 6

Trees and shrubs browsed at Shika and in the Zaria and Katsina areas

Species	Shika		Zaria	Katsina
	Cattle	Sheep		
Acacia sieberiana		xx		
A. seyal				xx
Adenodolichos paniculatus	xx			
Albizzia chevalieri		x		
Annona senegalensis			x	x
Anogeissus leiocarpus		x	x	x
Balanites aegyptiaca				x
Borassus aethiopicum	xx		x	
Bridelia ferruginea	xx	xx	x	
Burkea africana	x			
Butyrospermum paradoxum	xx			x
Capparis corymbosa	x			
Carissa edulis		xx	xx	xx
Combretum aculeatum	x			
C. ghasalense	x			x
C. glutinosum				xx
Commiphora pedunoculata				xx
Cussonia barteri	x			
Daniellia oliveri	x			
Detarium microcarpum		x		
Dichrostachys cinerea	xx	xx	xx	
Entada africana				x
Feretia apodanthera	xx	xx	xx	
Ficus spp.	x	x	x	
Grewia spp.	xx			

Species†	Shika		Zaria	Katsina
	Cattle	Sheep		
Guiera senegalensis	x		xx	xx
Hymenocardia acida				
Hyphaene thebaica				xx
Khaya senegalensis	x	x		
Lannea cf. schimperi		x		
Nauclea latifolia	x			
Ochna afzelii		x		
Parinari curatellifolia	x			
Parkia clappertoniana	xx			
Pavetta crassipes	x			
Piliostigma spp.	xx	xx	xx	xx
Prosopis africana				x
Pseudocedrela kotschyi	x			
Pterocarous erinaceous	x			
Sclerocarya birrea		xx		
Securinega virosa	x	x		
Sterculia setigera				x
Stereospermum kunthianum	xx			xx
Strychnos spinosa	x	xx		
Swartzia madagascariensis	x		x	
Tamarindus indica	x			x
Terminalia spp.	x			x
Vitex doniana	x	x		
Ximenia americana		x	x	
Ziziphus mucronata		x		xx

xx = Readily consumed species.

x = Occasionally consumed species.

† = Nomenclature according to Keay *et al.* (1954–68).

In pastoral herds

During the time analysis studies involving semi-nomadic and sedentary herds, browse species were recorded. Many species were eaten but since browsing is very much concentrated at the end of the dry season, browsing frequency was low for most plants. Several species are not touched for most of the year and eaten only when producing new leaves, fruits or flowers (*Combretum glutinosum, Capparis corymbosa* and *Acacia seyal*). The most important browse species for the Katsina and Zaria survey areas, have been tabulated in Table 6.

THE PALATABILITY OF NON-GRAMINACEOUS HERBS

Preferences of stock for non-graminaceous herbs were recorded in the same way as for woody plants. However, the records were much less detailed because the observation of the grazing of low and prostrate plants is much more difficult than that of shrubs (Table 7). In Shika, several recordings were done with

TABLE 7

Non-graminoid herbs consumed by stock at Shika, Zaria and Katsina

Species‡	Shika		Zaria	Katsina
	Cattle	Sheep†		
Aspilia spp.	xx	xx		
Borreria radiasa	xx			x
B. verticellata	x	xx	x	
Cassia tora		x	x	x
Cochlospermum tinctorium	xx			x
Cryptolepis sanguinolenta	xx			
Eriosema griseum	x	x		
Indigofera arrecta	x	xx		
I. pulchra	xx	xx		
Monechma ciliatum	x			
Nelsonia canescens	x		xx	x
Sida linifolia		xx		
S. rhombifolia	x			
Sesbania pachycarpa	x			
Stylosanthes mucronata	x	xx	x	x
Tridax procumbens			x	
Triumfetta pentandra	x		x	
Urena lobata	xx	x	x	
Vernonia purfurea	x			
Waltheria africana	x			
Zornia glochidiata			x	xx

†Excluding species listed in Table 8.
‡Nomenclature according to Keay *et al.* (1954–68).
xx = Readily consumed species.
 x = Occasionally consumed species.

TABLE 8

Non-graminoid herbs grazed by sheep in Shika†

Readily consumed	*Occasionally consumed‡*
Achyranthus aspera	Acalypha segethalis
Asparagus flagellaris	Acanthospermum hispidum
Clematis hirsuta	Amaranthus spinosus
Crotolaria retusa	Cuculigo pilosa
Euphorbia heterophylla	Evolvulus alsinoides
Schwenckia americana	Spigelia anthelmia
Sida acuta	Tephrosia linearis T. purpurea
Scoparia dulcis	Wissadula amplissima
Tephrosia bracteata	
Triumfetta dubia	

† = Excluding species listed in Table 7.
‡ = Nomenclature according to Keay *et al*. (1954–68).

sheep grazing pastures that were sown with *Stylosanthes humilis* and *S. guyanensis* and therefore contained a wide variety of fallow weeds. Many of these were so readily eaten that it might be worthwhile to try sheep grazing to control weeds in newly established pasture (see Tables 7 and 8).

DISCUSSION

The finding that maximum browsing occurred at the end of the dry season when grasses are of extremely low nutritive value often in short supply, while at the same time most shrubs produce new leaves and often fruits and flowers, has been confirmed by many studies in Africa and elsewhere both with domestic stock and wild ungulates. Browse in fistula samples from cattle grazing desert shrub in Nevada varied from 2 to 98 per cent and this variation was largely associated with the periodicity of grass herbage availability (Connor *et al*., 1975). Nearly a sole diet of browse was reported when cattle grazed pollarded Miombo woodland in Zambia with an estimated consumption of 3.2 per cent of body weight (Rees, 1974). Similarly, in semi-arid savannah in Tanzania, Payne and Macfarlane (1963) recorded browsing to grazing ratios of between 16 to 35 per cent and found a close correlation between the quantity of preferred species and browsing time. In environments with more available grass herbage, browsing by cattle is usually low and varies from 3 to 12 per cent (van Dyne and Heady, 1965, Blankenship and Quortrup, 1974).

Species selected for gramineae and sedges has been well-documented for the Sahel and Sudan zones in West Africa. In Table 9 the relative preferences of domestic stock for three groups of plants has been extracted from surveys done in Mauritania (Mosnier, 1961), Niger (Peyre de Fabreques, 1967) and Mali (Boudet, 1962). Avoided grasses comprised 14 per cent in Mauritania and 43

TABLE 9

The relative palatability of West African forage plants

Group of plants	Grasses and sedges			Trees and shrubs			Non-graminoid herbs		
Territory*	Niger	Mali	Maur.	Niger	Mali	Maur.	Niger	Mali	Maur.
Palatability score†									
High	26	6	14	10	1	5	5	2	10
Medium	36	21	31	20	16	16	13	5	11
Low	15	30	41	28	—	14	14	15	23
Zero	23	43	14	42	83	68	68	78	56
Total number of plants listed	115	200	71	74	147	63	275	257	140

*Niger (Peyre de Fabreques, 1967), Mali (Boudet, 1962), Mauritania (Mosnier, 1961).
†Scores in per cent of total plants listed.

per cent in Mali; in the Mali survey some 50 sedges were included in the graminoid group, most of which are little eaten by domestic stock. Low preference for sedges is also reported for some wild ungulates (Stewart and Stewart, 1970; Blankenship and Quortrup, 1974). Most low-growing fine-stemmed annual grasses belonging to the genera *Brachiaria, Chloris, Aristida, Digitaria* and *Eragrostis* are readily grazed while the perennial species within the same genera are much less palatable. The tall perennial tussock grasses in savannah with rainfall above 1000 mm are even less preferred by stock. *Andropogon* (except *A. gayanus*), *Hyparrhenia* (except *H. rufa*), *Loudetia* and *Schizachyrium* species which dominate these savannah are relatively unpalatable (Boudet, 1962, Peyre de Fabreques, 1967). However, when stock have little choice and graze these perennial grasses, grazing preferences differ little (Table 5). It is true that relative preferences of stock were based on the frequency ratios between grasses present and grazed and therefore probably much less accurate than methods involving fistulated animals (van Dyne and Heady, 1965; Bredon *et al.*, 1967). However, low preferential grazing may also be associated with the rather uniform growth forms of the dominant grasses, which during the recording period differed little in stages of maturity and leaf/stem ratio.

The wide spectrum of trees and shrubs that are included in diets of grazing livestock in the Sahel and Sudan zone is shown in Table 6 and 9. In this region between 20 to 60 per cent of the species are browsed to a variable extent. However, a comparison between species lists from different countries reveals marked differences in preferences ratings. For instance, Diallo (1973) listed 61 browse trees in Senegal, many of which are believed to be avoided by stock in other areas of the same ecological zone. While all Acacias are considered valuable browse species, Mosnier (1961) list some *Caparidaceae* and most *Combretaceae* as unpalatable as in contrast to data from Nigeria (Table 6), Senegal (Diallo, 1973) Niger (Bartha, 1970) and East Africa (Dougall *et al.*, 1964). Such comparisons bring out that animal preferences for browse vary widely with the season and the alternative vegetation so that the palatability of the flora can often not be predicted from past observations (Wilson, 1969).

Much less is known about preferences for the common species of the Guinea zone, where the arboreal cover is denser, more broad-leaved and less deciduous than further north. From the limited data presented here, it might be estimated that from the hundred or so species of that make up these savannahs, some forty are likely to be browsed — when occurring as shrubs. Many of these, however grow to trees offering little foliage within reach of grazing stock. The Guinea zone is very close to the Miombo woodland in respect of taxons and physiognomy and the experiments with cattle grazing in pollarded woodland in Zambia are of special interest (Rees, 1974).

Non-graminoid herbs are generally less preferred than grasses and shrubs — between 60 and 90 species are reported to be grazed (Table 9), while Dougall *et al.* (1964) in East Africa recorded only 24 herbs eaten by ruminants.

Palatability scores for species differ greatly between workers so that it is no surprise that between 70 and 80 per cent of the species listed as grazed in Table 7 were considered unpalatable by either Boudet (1962), Mosnier (1961) or Peyre de Fabreques (1967). Bartha (1970) working north of Niamey listed 60 grazed dicotyledons; 51 are also quoted by Peyre de Fabreques (1967) in the Zinder area in Niger, of which only 30 were recorded as acceptable to stock.

This review of largely West African literature together with the data from Nigeria show that, although it is evident that domestic ruminants select their food from a wide range of plants, recorded preferences for fodder species differ greatly. This is not surprising as it is known that these preferences are affected by climate, soil and the species composition of the available range vegetation and also vary with the seasonal fluctuations in fodder quantity and quality (Heady, 1964). It is also obvious that palatability of plants and resulting food preferences of animals are difficult to measure accurately, thus further research should include the use of precise techniques for determining diet composition and nutritive value within well-defined grazing and management systems.

ACKNOWLEDGEMENTS

The author is indebted to Mr S. O. Magaji, Mr M. Oche and Mr M. Ayuba for their assistance with field work in Katsina and Zaria and to staff of the Grassland Section for help with behaviour studies in Shika. Thanks are due to the Director of the Institute for Agricultural Research for the permission to publish.

REFERENCES

BARTHA, R. (1970) Fodder plants in the Sahel zone of Africa. (IFO-Institut, Munich).

BLANKENSHIP, L. H. and QUORTRUP, S. A. (1974) Resource management on a Kenya ranch. *South African Wildlife Association Journal*, Vol. 4, pp. 185–90.

BOUDET, G. (1962) Etude botanique et agrostologique de la haute valle du Niger. *Rev. Elev. Med. Vet. Pays trop.*, Vol. 15, pp. 75–105.

BREDON, R. M., TORELL, D. T. and MARSHALE, B. (1967) Measurement of selective grazing of tropical pastures using esophageal fistulated steers. *Journal of Range Management*, Vol. 20, pp. 317–20.

CONNER, J. M., BOHMAN, V. R., LESPERANCE, A. L. and KINSINGER, F. E. (1963) Nutritive evaluation of summer forage with cattle. *Journal of Animal Science*, Vol. 22, pp. 961–9.

DIALLO, A. K. (1973) Problems posés par l'utilisation des especes ligneuses dans l'alimentation des animaux domestiques Senegalais en zone d'elevage extensif. *Association for the Advancement of Agricultural Sciences in Africa*, Vol. 1, pp. 45–55.

DOUGALL, H. W., DRYSDALE, V. M. and GLOVER, P. E. (1964) The chemical composition of Kenya browse and pasture herbage. *East African Wildlife Journal*, Vol. 2, pp. 86–121.

HAGGAR, R. J. (1968) Grazing behaviour of fulani cattle at Shika, Nigeria. *Tropical Agriculture*, Vol. 45, pp. 179–85.

HEADY, H. F. (1964) Palatability of herbage and animal preference. *Journal of Range Management*, Vol. 17, pp. 76–82.

KEAY, R. W. J., HEPPER, F. N. and ALSTON, A. H. G. (1954–68) *Flora of West tropical Africa* 2nd ed., (Crown Agents, London).

MOSNIER, M. (1961) Pasturages naturels sahels: Region de Keadi (Mauritanie). (IEMVPT, Maison Alfort, France).

RAAY, J. G. T. VAN and DE LEEUW, P. N. (1970) The importance of crop residues as fodder: A resource analysis in Katsina province, Nigeria. *Tyd. econ. Soc. Geogr.*, Vol. 40, pp. 137–47.

RAAY, J. G. T. VAN and DE LEEUW, P. N. (1974) Fodder resources and grazing management in a savannah environment: an ecosystem approach. (Institute of Social Sciences, The Hague) p. 29.

REES, W. A. (1974) Preliminary studies into bush utilization by cattle in Zambia. *Journal of Applied Ecology*, Vol. 11, pp. 207–14.

ROSE-INNES, R. R. and MABEY, G. L. (1964) Studies on browse plants in Ghana. 3. Browse ingestion ratios. *Empire Journal of Experimental Agriculture*, Vol. 32, pp. 180–90.

STEWART, D. R. M. and STEWART, J. (1971) Comparative food preferences of five East African ungulates at different season. *Scientific management of animal and plant communities for conservation.* (Blackwells) pp. 351–66.

WILSON, A. D. (1969) A review of browse in nutrition of grazing animals. *Journal of Range Management*, Vol. 22, pp. 23–8.

VAN DYNE, G. M. and HEADY, H. F. (1965) Botanical composition of sheep and cattle diets on a mature annual range. *Hilgardia*, Vol. 36, pp. 465–92.

14

THE USE OF FAECAL ANALYSES IN STUDYING FOOD HABITS OF WEST AFRICAN UNGULATES

Chris Geerling

Department of Forest Resources Management, University of Ibadan, Nigeria†

INTRODUCTION

The epidermis of plant leaves shows characteristic and often specific patterns in the arrangement and shape of cells, stomata, silica bodies, hairs and prickles. The epidermal characteristics of grasses, which have taxonomical value have received much attention (Metcalfe, 1960).

The epidermis, or at least its cuticule, which forms the protective layer for the plant, is very resistant to chemical and mechanical processes and survives the digestive processes of herbivores in recognizable form. The epidermis fragments found in herbivores faeces can thus serve to identify plant species eaten by the animal.

Following various authors who describe the use of faecal analyses in determining food habits of free-ranging domesticated animals in temperate zones, Stewart (1963) made an extensive study of its use in determining food habits of wild herbivores in East Africa.

In order to test the method under West African conditions I analysed droppings of roan antelope (*Hippotragus equinus* Desmarest) and Western Hartebeest (*Alcelaphus buselaphus* Pallas) — two large antelopes which apparently share the same habitat in West Africa.

STUDY AREA

The droppings were collected in the central area of Kainji National Park, Kwara State, Nigeria. This Park is situated between the river Niger and Benin Republic, between latitude 9° and 11° N. and longitude 3° and 5° E.

The landscape consists of rolling hill country, underlain by metamorphic

‡Present address: FAO Forestry Office (Range Management), Ecole de Faune, Garoua, Cameroun Republic.

rocks, mostly migmatites, of the Basement Complex, with quartzite ridges in the central part of the park. The Oli river, which flows from west to east through the southern part of the park, drains the larger part of the park.

The vegetation of the central area consists of *Afzelia africana* Don woodland on upland sites, *Burkea-Terminalia avicennioides* Guill. and Perr. savannah woodland on the lower slopes and *Isoberlinia* woodlands adjacent to the quartzite ridges (Geerling, 1976).

METHODS

Faecal analyses based on the epidermal characteristics of grasses involves two steps: the preparation of a reference collection of epidermal slides, as most of the epidermal characteristics of West African grasses have not yet been described, and the actual analyses of the droppings.

The reference collection

A survey of the grasses of the study area was made from September–November 1972. Flowering plants were collected for identification and leaves of the same specimens were collected and conserved in alcohol.

Epidermal slides were prepared of the middle of the leave blade, either by scraping with a razor blade in a commercial bleach solution, or by boiling part of the leave in concentrated nitric acid (Metcalfe, 1960; Stewart, 1963). Specimens prepared by the latter method are cleaner but more difficult to handle than those prepared by scraping. Temporary slides were prepared, using a water–glycerine mixture. Microphotos were taken by ultra-violet light. A key for about 50 of the common grass species was prepared — for both the abaxial and the adaxial sides of the leaf blade.

The analyses of the faeces

Fresh droppings of roan antelope and hartebeest were collected in the morning, on a sandy track which had been cleared of old droppings the evening before. The droppings were collected on six occasions, from late December 1972 (early dry season) till early June 1973 (early rainy season) at about 5 week intervals. From each species 10 pellet groups were collected and conserved in alcohol.

For the analyses, the pellets were crushed and mixed, and a sample was boiled in concentrated nitric acid. After boiling the nitric acid was removed by repeatedly diluting with water, sedimenting and pouring off. Sedimenting was much less laborious than centrifuging, especially when working with large series. For the analyses samples were pipetted onto slides with a water–glycerine mixture.

TABLE 1

Percentage of grass epidermis fragments in roan antelope and hartebeest faeces

Collection period	Roan antelope	Hartebeest	Difference
Dec.	88.9	97.8	Highly significant ($p = 0.01$)
Jan.	72.8	77.3	Not significant
Feb.	87.0	91.1	Significant ($p = 0.05$)
Mar.	57.7	78.5	Highly significant
May	96.1	99.3	Significant
Jun.	99.2	99.6	Not significant
Mean	83.4	90.4	Significant

RESULTS

The analyses of roan antelope and hartebeest droppings had to be limited to the relative frequency of occurrence of grass and forb epidermis fragments because of difficulties in identifying the often closely related grass species.

The percentage of grass epidermis fragments in roan antelope and hartebeest droppings is given in Table 1. Forb epidermis fragments make up the difference to 100 per cent. In each collecting period, 10 pellet groups of each species were collected and analysed.

Roan antelope faeces showed a significantly higher portion of forb epidermis fragments than hartebeest faeces. The ratio forb/grass epidermis fragments showed considerable variation for the different collecting periods. This is probably related to the phenology of the early burnt savannah woodland vegetation of the study area (Geerling, in the press).

DISCUSSION

Faecal analyses gives the relative frequencies of diet components for which epidermal fragments remain recognizable in the faeces.

The method, however, has its limits:

(1) The relative frequency of diet component does not permit assessing the importance of a diet component for the animal, because some diet components do not have a recognizable epidermis, e.g. fruits, and different plants may have different nutritive values and the relative frequency of occurrence in the faeces is not necessarily a measure of the relative importance of the diet component.

(2) The method is laborious, as it demands an intensive study of the vegetation of the area concerned (relative frequency or biomass of potential food species) and the preparation of a reference collection of epidermal slides and a thorough knowledge of the characteristics of each species.

(3) The Andropogonae are the predominant grasses of the West African Guinea and Sudan (broad-leaf, leguminous and combretaceous) savannah woodlands, both in number of species and in biomass. The high number of often closely related *Hyparrhenia* and *Andropogon* species make the identification of small epidermis fragments difficult if not impossible.

APPLICABILITY OF THE METHOD

(1) For management purposes. If the necessary equipment (microscope, with 100× magnification, slides and covers, test-tubes and concentrated nitric acid) is available, faecal analyses can be used to assess the proportion of browse and/or grass taken, and to detect changes in the diet. This may give useful information in planning the burning regime, in order to produce a flush of either grass or browse in the right place at the right time.

(2) In research, under general West African conditions as described above, faecal analyses is not likely to produce valuable information in relation to the amount of work to be done. The method can be used, as outlined above, for the assessment of the proportion of forbs and grasses taken. Little or no work has been published on the epidermal characteristics of West African woody plants, but the number of species in a given area is usually limited and a reference collection will be relatively easy to make.

The method can also be used on flood-plains (rare in West African wildlife areas), where grass species are both relatively low in number and taxonomically more diverse than in savannah woodland.

ACKNOWLEDGEMENTS

Mr I. Iyamabo, Director of the Federal Department of Forest Research, generously provided transport for the field work. Mr N. O. Dunsin assisted in both the field work and the analyses.

REFERENCES

GEERLING, C. (1976) *1:50 000 vegetation map of Borgu Game Reserve.* (FAO Working Document. FAO Forestry Department FAO, Rome.)

METCALFE, C. R. (1960) *Anatomy of the Monocotyledons, I. The Graminees.* (Oxford University Press.)

STEWART, D. R. M. (1963) The determination of food preferences of wild herbivores by faecal analysis. (Ph.D.Thesis, University of Nairobi, Kenya.)

15

HABITAT SELECTION AND ECONOMIC IMPORTANCE OF RODENTS IN MOOR PLANTATION, IBADAN, NIGERIA

Oluwadare Funmilayo

Department of Agricultural Biology,
University of Ibadan, Nigeria

and

Modupe Akande

Institute of Agricultural Research and Training,
University of Ife, Moor Plantation, Nigeria

INTRODUCTION

Rodents are one of the most successful groups of animals. The magnitude of the success of rodents in the terrestrial habitats of tropical regions is indicated by the numerical abundance of the group, the large number of species and the wide morphological variability in the group to suit existence in the soil for fossorial species, on land for terrestrial species, on tree tops for arboreal species and inside buildings for commensal forms.

Notes on the identification of the rodent species in West Africa have been made by Booth (1960) and Rosevear (1969). Everard (1966, 1968) dealt mainly with rodent damage to crop plants but his reports also contain useful information on the distribution of the rodent pest species in his study area. Accurate knowledge of the distribution of a species is needed in the effective management and control of that species.

The present study examines the choice of living space by rodents in Moor Plantation, Ibadan. The information on damage is supplied to indicate some of the biological influences of rodents in both their natural and man-made environments. Rodent damage is also of considerable economic importance agriculturally, domestically and industrially. The use of rodent flesh for meat (Ajayi, 1974) is also an important factor in the local economy and diet.

MATERIALS AND METHODS

The investigations were carried out in Moor Plantation, Ibadan. Some of the physical features of this area have been described by Funmilayo (1975). The site lies within the lowland moist forest zone, the highest elevation in Moor Plantation being 183 m above sea level. The average annual rainfall is just over 1000 mm, the rain falling mainly from March to early November with a short dry period in August. During the peak dry season, from late November to February, the dry and dessicating Harmattan wind blows from the north-east.

The study area, originally occupied by high forest, whose relics can now be seen only in a few areas, has suffered persistent human disturbances from agriculture, house and road building since 1905 which has produced the present mosaic of woodlands, bush fallows, arable lands, lawns and buildings. There are four distinct habitats in the study area — woodlands, arable lands, swampy stream banks and buildings.

The woodland areas included the relics of the original forest, old (more than three years) bush fallows, planted woodlands which include cocoa, rubber and coconut palm plantations with scattered plantain and banana trees. A thick ground cover of herbs and weeds with scattered accumulated debris of broken stems and dropped leaves is a common feature of the woodland areas.

Arable lands are open farmlands used for experimental rotation cropping. Parts of the arable lands are planted with various food crops seasonally while the uncultivated parts grew a dense, often matted cover of weeds and grasses.

The 'Alagbon' stream flows west–east across the site. The soil on both sides of the stream was swampy and covered with a thick layer of dead grass stems and leaves. The vegetation was almost impenetrable and consisted predominantly of elephant grass (*Pennisetum purpureum*), and shrubs with climbing and scrambling plants.

Buildings in the study area included residential houses, office blocks, a seed store, a poultry house and a rabbitry.

Rats and mice were captured mainly by using snap traps baited with ripe oil-palm fruits. Some rats and mice were also caught by hand during the search for rodents and their nests in holes, accumulated debris and rubbish dumps. All giant rats were obtained in live traps. Tree squirrels were obtained with steel wire snares. Nests and burrows were assigned only to species killed or consistently found in such nests and burrows. The distribution of each rodent species was determined from the capture records and visual observations. The types of damage observed in the study area were recorded and the rodent species causing damage were determined by trapping, visual observations, and in some cases, by the examination of stomach contents.

RESULTS

Distribution of rodents in the study area

The distribution of rodent species in the study area is shown in Table 1. 113 rodents belonging to 12 species were captured in woodlands, 147 rodents belonging to 10 species in arable lands, 43 rodents belonging to 8 species in

TABLE 1

The distribution of rodents in Moor Plantation, Ibadan

Species of rodents	Woodlands		Arable lands		Swampy stream banks		Inside buildings		All habitats		
	M	F	M	F	M	F	M	F	M	F	M + F
Family: Sciuridae											
F. anerythrus	7	9							7	9	16
H. gambianus	3	2							3	2	5
P. strangeri	3								3		3
Family: Cricetidae											
T. kempi			4	3					4	3	7
C. gambianus	4	6	1	1			4	6	9	13	22
Family: Muridae											
R. rattus	1	1	3	1	2	5	19	39	25	46	71
A. niloticus			7	8	2	3			9	11	20
L. sikapusi	13	12	13	5	11	7			37	24	61
D. incomtus	3	1	12	3	2				17	4	21
L. striatus	5	5	2	1	1	2			8	8	16
T. rutilans	2	1							2	1	3
U. foxi	4	2	17	10	2			1	23	13	36
P. tullbergi	5	4			1	1			6	5	11
M. natalensis	8	4	27	16	3	1	6	5	44	26	70
M. musculoides	3	2	10	3					13	5	18
S. longicaudatus	2	1							2	1	3
Family: Thryonomidae											
T. swinderianus					2	2			2	2	4
All rodent species	63	50	96	51	24	19	29	51	212	171	383

swampy stream banks and 80 rodents belonging to 4 species inside buildings. The distribution of individual species is discussed below.

Family: Sciuridae

Redless striped squirrel — *Funisciurus anerythrus* Thomas (1890).

This diurnal tree squirrel was captured only in woodlands. The rounded nest, with a single exit, is made of finely-shredded, dried tree bark and is usually placed on shrubs, small trees or forest tangled vegetation at a maximum distance of 4 m above soil level. This species forages mainly on trees but may occasionally use the soil surface to cross roads or other such discontinuity in the vegetation.

Gambian sun squirrel — *Heliosciurus gambianus* Ogilby (1835).

This is also a diurnal tree squirrel restricted to woodlands. The nests of this species were placed among living and dead oil-palm tree leaf petioles and in holes in tree trunks. This species is almost restricted to tree tops except for the very occasional visits to obtain fruits on dislodged plants like pawpaw, plantain and banana.

Giant forest squirrel — *Protoxerus strangeri* Waterhouse (1843).

This is also an exclusively high forest species which, unlike *F. anerythrus* and *H. gambianus*, is absent from bush fallows but found only in the upper storey of tall trees in the mature forests. Nests are placed in similar locations to *H. gambianus*. *P. strangeri* is very seldom seen foraging on the ground surface.

Family: Cricetidae

Kemp's gerbil — *Tatera kempi* Wroughton (1906)
This is a burrowing, nocturnal species that is restricted to the soft soils of cultivated lands. Commonly, there are no external indications, like exits, of the shallow burrows, from which many of the rats escape as the burrows are punctured during cultivation. The presence of this species can be recognized from its characteristic feeding habit of gnawing fairly old and weak maize, rice and sorghum stems at about 10 cm above soil level.

Giant rat or Gambian rat — *Cricetomys gambianus* Waterhouse (1840).

This is also called the pouched rat because of its enormous cheek pouches in which it carries away yam tubers, maize cobs, oil-palm fruits and other food materials to be stored in its burrows and consumed at its own leisure. This large rat inhabits both simple and elaborately designed burrows, made or acquired by it in all soil types, inside dead wood and even inside houses. Seven specimens were captured inside an office where they were defecating daily inside desk drawers. This species is a fast runner, climbs efficiently and is also able to swim. It is hunted for food in most parts of its range though it is tabooed in a few areas. *C. gambianus* was formerly included in the family Muridae but Rosevear (1969) placed it in the family Cricetidae.

Family: Muridae Gray

Black/grey rat — *Rattus rattus rattus* Linnaeus (1758).

This is the most widely distributed and the most abundant rodent species in the study area. It was captured in woodland, arable land, swamps and inside buildings. Few specimens were found in woodlands and the highest numbers were obtained inside buildings and it is, therefore, a commensal or domestic species. This rat is an efficient burrower, climber and jumper, though not really a fast runner. It nests mainly in house roofs and in holes. It is mainly nocturnal but it has been observed gnawing the cardboard and celotex roof ceilings of offices during the day. It occurs mainly in two colour forms, the black/grey form (*R. r. rattus*) and the brown variety, (*R. r. alexandrinus* Geoffroy, 1803) though a few albino forms are found. The brown form (*R. r. alexandrinus*) has often been wrongly identified as *Rattus norvegicus*. Only one record (Buxton, 1936) has indicated the presence of *R. norvegicus* in Nigeria and only in the port city of Lagos.

Nile harsh-furred rat or Kusus rat — *Arvicanthis niloticus* Desmarest (1822).

This rat is absent from forests but it is a typical rat of grassy and weedy open lands and therefore one of the commonest rats on farmlands. It makes characteristically clean, wide runways through the vegetation. It is a non-burrowing, non-climbing, but very fast running rat. The round nest, composed of dried grass is usually placed unconcealed among weeds on the soil surface in the wet season but in the dry season nests are placed in holes, under wood, stones or rubbish dumps.

Rufous-bellied rat — *Lophuromys sikapusi* Temminck (1853).

This is a very common rat in forest clearings, littered and weedy floors of planted woodlands, grassy and weedy arable lands and damp areas. It is absent from primary forests except at its open edges. It is a nocturnal, non-burrowing, non-climbing, sluggish rat, relying for protection only on its ability to conceal itself or quickly retreat into any available burrow.

Shaggy rat — *Dasymys incomtus* Sundevall (1847).

This species was captured in woodland, arable land and swamps. It is a nocturnal, semi-aquatic, burrowing rat. The nest, a big ball of grass, is usually placed in tangled vegetation on the opening of a short blind burrow into which the rat usually, but not always, retreats when molested. This makes it easy to obtain the rats by digging it out of the short burrow.

Spotted grass mouse — *Lemniscomys striatus* Linnaeus (1758).

This species was captured in woodland, arable land and swamps. The nests and runways are very similar to, but smaller, than those of *A. niloticus*, with which it shares exactly the same habitats and habits. It is not a species of primary forest but occurs only at the cleared edges of forests. It could very easily be confused with a similar (the three-striped mouse, *Hybomys trivirgatus* Temminck, 1853), but exclusively forest dweller, of about the same size.

Shining thicket rat — *Thamnomys rutilans* Peters (1858).

This is a nocturnal, arboreal and exclusively forest species. The rounded nest of leaves is placed in a tangle of vegetation or in shrubs and trees at a maximum of 3 m from ground level. The rat lies in the nest in such a way that it can see the outside from the single exit, from which it quickly escapes, usually before it can be captured. The rat forages extensively on the ground surface.

Fox brush-furred rat — *Uranomys foxi* Thomas (1912).

This nocturnal species was captured in woodland, arable land and swamps but it was most abundant in cultivated lands. It occasionally enters farm buildings.

Tullberg's rat — *Praomys (Rattus) tullbergi* Thomas (1894).

This rat was captured mainly in woodlands and is an almost exclusively forest species. It is the most abundant rat in primary forests and cocoa plantations. It builds its nests in tree hollows, in rotten banana and plantain trunks and under fallen trees. It is an efficient climber and a fast runner. It forages a lot, at night, on the ground surface, without making any clearly defined runway.

Multimammate rat — *Mastomys natalensis* Smith (1834).

This species was captured in all parts of the study area. It is the most abundant rat on cultivated lands in all parts of the Western State of Nigeria. It is not as successful as *R. rattus* in its association with human dwellings, being restricted mainly to houses built very close to farmlands from which the rats quickly invade the houses and settle in the interior. Once inside the house they are able to manipulate their way around almost as easily as cockroaches. The nesting habits are similar to those of *R. rattus* except that on farmlands, *M. natalensis* may become gregarious in shallow tunnels.

Long-tailed target rat — *Stochomys longicaudatus* Tullberg (1893).

This is a rare nocturnal species of which only three specimens were captured in the forest. It nests in holes in trees and is a good climber.

Pigmy mouse — *Mus (Leggada) musculoides* Temminck (1853).

This species was captured in woodland and arable land. It is mainly a species of farmlands, grassy lawns and open forest areas. The very small nest, made of grass, is usually placed under stones, wood and other such covers. It can often be seen crossing roads at night or pursuing insects attracted by light. It often wanders into buildings but does not live inside buildings.

Family: Thryonomidae

Cane rat or Grass cutter — *Thryonomys swinderianus* Temminck (1827).

This is the largest terrestrial rodent in West Africa. The trapping techniques used in this study were unsuitable for this species which was therefore not captured. However, four specimens were shot by a hunting party along the swampy stream banks. This species is absent from primary forests but common in degraded forests and farmlands, particularly damp areas with thick scrub of grasses (e.g. *Pennisetum purpureum, Panicum maximum*) and weeds (*Aspilia* sp., *Eupartorium odoratum*). It can come very close to human dwellings located within its habitat especially where maize or rice is planted as a garden crop.

ECONOMIC IMPORTANCE OF RODENTS IN THE STUDY AREA

The various types of rodent damage in the Western State of Nigeria have been documented in previous reports (Everard, 1966, 1968; Funmilayo, 1973; Funmilayo and Akande, 1974a, 1974b). A summary of the nuisance activities of rodents and other aspects of the economy of rodents are discussed here under each rodent family captured in the study area.

Family: Sciuridae

Each of the three species in this family consumed oil-palm fruits, pawpaw fruits, cocoa beans, maize grains, plantain and banana fruits removed from upright and fallen plants. Tree squirrels do not consume or damage post-harvested fruits. All the three species are eaten by the local populace and there are no known tabooes attached to the eating of squirrel flesh.

Family: Cricetidae

Kemp's gerbil (*T. kempi*) is known to cut and fell young and weak fairly old maize, rice and sorghum stems, consume the inner tissues of the stem and the grains. Other types of damage caused by this species include the removal of planted seeds, cutting of aerial shoots of yam and consumption of underground swollen roots and tubers. The gerbil is eaten in many parts of its range by the local people who capture it.

The giant rat (*C. gambianus*) removes matured cobs on upright, deformed, leaning and fallen maize stems. It cuts the aerial shoots of yam and also exposes and consumes yam setts and tubers, cassava roots and other underground swollen roots and stems. It consumes ripe fruits of oil palm, cocoa, pawpaw, banana and plantain. It has been observed to block the sewage system in houses in order to build its nest and rear its young in the soakaway pit. The giant rat is tabooed in many localities and the increasing incursion of the rat into human dwellings has tended to lower the social acceptability of its meat.

Family: Muridae

These small rats and mice are generally responsible for the removal of planted seeds on farms and gardens. They also cut young rice, maize, cowpea and groundnut stems. All rats and mice consume oil-palm fruits, many consume other seeds and fruits on the ground and in the soil. The climbing black/grey rat, *R. rattus*, and the multimammate rat, *M. natalensis*, also consume maize grains on upright plants.

There is scarcely anything in a house, including human beings, that the common black/grey rat, *R. rattus*, and to some extent the multimammate rat, *M. natalensis*, will not bite. These rats are known to bite human nails, hairs, toes, fingers and other parts of the body at night especially in houses where the rats occur in plague proportions and people are overcrowded. The rats will consume all food items and gnaw through chicken wire, lead pipes, wood, plastic and rubber. The black/grey rat was found to consume its own babies, other rats, rabbit kittens and bird chicks.

Most small rats and mice, except the commensal species and the diminutive pigmy mouse, *M. musculoides*, are eaten by the local populace, especially children. Some are also used in herbal concoctions and 'juju'.

Family: Thryonomidae

Cane rats damage food crops, mainly maize, rice, sorghum, cassava and sugar cane in the many parts of the study area. Generally, the cane rat is the most notorious rodent pest in the Western State of Nigeria. However, the soft, tender and sweet cane rat flesh is very popular and is regarded by both local people and foreigners as a delicacy.

DISCUSSION

The distribution of rodents in the study area have been described from trapping records and visual observations. Trapping records are notoriously difficult to interpret particularly when there are no physical barriers to prevent the free movements of animals from one part of the study area to the other. Another obvious difficulty in the interpretation of capture data is how to determine where a rodent lives as distinct from where it was captured, or which animal made a burrow or nest as distinct from the rodent, perhaps a fugitive, that was captured in it. Or, for instance, should an animal be regarded diurnal just because it was captured during the day even though it might have been frightened out of its nest or burrow by a predator or a disturbance (e.g. fire or flood) in its habitat? These limitations of trapping data as a means of determining distribution patterns and of assigning rodents to burrows and nests meant that trapping data must be corroborated by visual observations which must remain consistent over a long period.

Most of the rodent species captured in the study area, through their feeding and nesting habits, damage some components of the habitats that are of economic interests to man. Rodents destroy crop plants and also compete directly with man for both wild and cultivated fruits. Most damage is below the level of economic injury but recent results have shown that in South-Western Nigeria in particular, rodent damage may seriously impair man's efforts to produce food (Everard, 1966, 1968; Funmilayo, 1973; Funmilayo and Akande, 1974b) and even animal protein (Funmilayo and Akande, 1974a).

Commensal rats consume foodstuffs and damage a wide range of household goods by gnawing or consuming them. Damage caused by rats in homes, stores and commercial premises is now considered to be a serious economic problem in Nigeria. Damage apart, commensal rats are undesirable companions of man, not only because they carry deadly human disease agents (Akande and Funmilayo, 1974) but also because they disturb peace and rest and foul food and the home environment with their urine, faeces, tail and feet marks.

ACKNOWLEDGEMENTS

This work was carried out with funds and facilities provided by the Institute of Agricultural Research and Training, University of Ife, Ibadan for which we are grateful to the authorities. We would also like to thank our colleagues who have allowed the trapping of rats in their houses and offices.

REFERENCES

AJAYI, S. S. (1974) Giant rats for meat and some taboos. *Oryx,* Vol. 12, pp. 379–80.

AKANDE, M. and FUNMILAYO, O. (1974) Some common diseases that can be transmitted from rodent to man. Research Bulletin No. 4. (Institute of Agricultural Research and Training, University of Ife, Ibadan, Nigeria.)

BOOTH, A. H. (1960) *Small mammals of West Africa.* (West African Nature Handbooks, Longmans, London.)

BUXTON, P. A. (1936) Breeding rates of domestic rats trapped in Lagos, Nigeria and certain other countries. *Journal of Animal Ecology*, Vol. 5, pp. 53–65.

EVERARD, C. O. R. (1966) Reports on some aspects of rodent damage to maize in the Western Region of Nigeria. (Research Division, Ministry of Agriculture and Natural Resources, Moor Plantation, Ibadan, Nigeria.)

EVERARD, C. O. R. (1968) A report on rodent and other vertebrate pests of cocoa in Western Nigeria. (Research Division, Ministry of Agriculture and Natural Resources, Moor Plantation, Ibadan, Nigeria.)

FUNMILAYO, O. (1973) A general survey of the incidences and control methods of vertebrate pest of crop plants in the Western State of Nigeria. Research Bulletin No. 1. (Institute of Agricultural Research and Training, University of Ife, Ibadan, Nigeria.)

FUNMILAYO, O. (1975) The ecology and economic importance of the rufous-bellied rat, *Lophuromys sikapusi sikapusi* in Ibadan, Nigeria. *Nigerian Journal of Plant Protein*, Vol. 1, pp. 17–22.

FUNMILAYO, O. and AKANDE, M. (1974a) The black/grey rat, *Rattus rattus rattus*, as a pest in a rabbitry in Ibadan. *Nigerian Journal of Forestry*, Vol. 4, No. 1, pp. 24–8.

FUNMILAYO, O. and AKANDE, M. (1974b) The ecology, economic impact and control of vertebrate pests of upland rice in the Western State of Nigeria. Research Bulletin No. 5. (Institute of Agricultural Research and Training, University of Ife, Ibadan, Nigeria.)

ROSEVEAR, D. R. (1969) *The rodents of West Africa*. (British Museum (Nat. Hist.), London.)

16

BODY WEIGHT, DIET AND REPRODUCTION OF RATS AND MICE IN THE FOREST ZONES OF SOUTH-WESTERN NIGERIA

Oluwadare Funmilayo

Department of Agricultural Biology,
University of Ibadan, Nigeria

and

Modupe Akande

Institute of Agricultural Research and Training,
University of Ife, Nigeria

INTRODUCTION

Investigations dealing with the biology and ecology of rats and mice in Nigeria are few (Rosevear, 1953, 1969; Anadu, 1974). Local rats and mice have been studied more because of their agricultural pest status (Everard, 1966, 1968; Funmilayo, 1973, 1975) and their medical importance (Buxton, 1936). Various rat control strategies have been developed (Chitty and Southern, 1954; Mosby, 1963) but in spite of these, rats continue to ruin crop plants and stored products (Funmilayo, 1973) and to endanger human health (Akande and Funmilayo, 1974). The poor success of previous attempts to control rats is probably a reflection of our inadequate knowledge of the basic factors which determine the population density and diet of rats and there is, therefore, a need to build up a large body of knowledge on rat ecology.

The data that are often needed for planning effective rat control strategies include those on the body weight, the weight of stomach contents, the composition of the natural diet and the seasonal trends in population numbers. The body weight of a rat is needed to determine the amount of a poison needed to kill the rat and the stomach content weight is a measure of the amount of food consumed by a rat at one feeding. Therefore, if one dose of poison is to be used to kill a rat, the food bait and the selected poison must be mixed in such a proportion that a quantity of the active ingredient of the poison required to kill the rat must be present per quantity of the poisoned-bait that is equivalent to or slightly less than the stomach content weight of the target rat. The natural diet

136

of the rat gives an indication of the suitable food materials to be used to mix poisoned bait. The data on the reproduction of rats in their natural habitat are necessary in order to forecast population trends and therefore when damage by rats, which is usually a function of population numbers, is likely to be most severe.

When rats cause damage there is usually a need for quick action which does not permit extensive data gathering before control is applied. Most of the data on the ecology of the pest species can therefore be obtained only from previous records. The present report supplies some of the hitherto absent, but necessary, data needed for the management of rats and mice in the study area.

MATERIALS AND METHODS

Rats and mice were captured in the Ekiti, Ijesha, Akure and Ijebu districts, which are all in the high forest areas of south-western Nigeria. Local meteorological records show a minimum annual rainfall of approximately 1000 mm for the study area. The rain falls mainly from March to early November with a short dry spell in August. The dessiccating and dusty Harmattan wind blows from the north-east during the dry season from late November to February. The surface vegetation dries up in the dry season and then it is burnt — except in the high forests and in areas with crop plants.

Rats and mice were captured mainly during rodent control operations of farmlands and inside building at irregular intervals between March 1971 and April 1975. Rats were also captured in high forests in the Ibadan area. The date of capture, sex, body weight and stomach content weight of each rat were recorded. Stomach contents were hardened in 40 per cent formalin for 24 hours and then placed in the following categories: oil-palm fruit, seeds, fruits, plant materials, adult insects, insect larvae, earthworms, myriapods and molluscs.

The number of embryos in each pregnant female was recorded. Body sizes and the reproductive organs were used as criteria for classifying rats into adults and juveniles. Generally, a rat weighing less than 50 per cent of the mean body weight for the species was classified as a juvenile. The mean body weight was calculated from the body weight measurements of adults.

RESULTS

The body and stomach content weight of rats and mice captured in the study area are shown in Table 1. Only three species, the black/grey rat, *Rattus rattus rattus* Linnaeus (1758), the shaggy rat, *Dasymys incomtus* Sundevall (1847), and the Nile harsh-furred rat, *Arvicanthis niloticus* Desmarest (1822), have a body weight of 100 g or slightly more. Four species, the multimammate rat, *Mastomys natalensis* Smith (1834), the spotted grass mouse, *Lemniscomys striatus* Linnaeus (1758), the rufous-bellied rat, *Lophuromys sikapusi*

TABLE 1

Body weight and stomach content weight of rats and mice

Species (No. of specimens in brackets)	Mean (+S.E.)			
	Body weight (g)		Stomach content weight (g)	
	M	F	M	F
Rattus rattus (171)	95.1±15.0	112.0±14.0	3.0±0.3	2.1±0.4
Dasymys incomtus (21)	110.6± 4.5	88.0±10.6	3.3±0.4	3.2±0.5
Mastomys natalensis (120)	56.6± 2.9	52.4± 5.2	1.5±0.1	2.1±0.3
Arvicanthis niloticus (40)	121.0±21.9	97.8±14.8	3.5±0.9	4.4±1.1
Uranomys foxi (56)	39.3± 2.4	35.3± 2.6	1.0±0.1	1.0±0.2
Mus musculoides (28)	7.5± 0.5	7.8± 1.2	0.4±0.1	1.0±0.1
Lemniscomys striatus (36)	51.8± 2.9	53.7± 2.4	2.5±0.4	2.6±0.4
Praomys Tullbergi (41)	40.3± 4.4	34.3± 1.9	1.9±0.2	1.5±0.2
Lophuromys sikapusi (61)	81.6± 3.0	67.9± 4.7	2.7±0.2	2.8±0.3
Stochomys longicaudatus (3)	30.0		2.0	
Hybomys trivirgatus (11)	54.5± 2.6	51.2± 2.4	2.8±0.3	2.6±0.2
Thamnomys rutilans (3)	42.0	38.0	2.5	2.5

TABLE 2

Qualitative composition of the stomach contents of rats and mice

Species (No. of specimens in brackets)	% number of stomachs containing			
	Plant materials	Insects	Earthworms	Other invertebrates
Rattus rattus (171)	88.9	33.3		
Dasymys incomtus (21)	84.2	7.0	5.3	
Mastomys natalensis (120)	74.0	52.0	4.0	
Arvicanthis niloticus (40)	91.3	39.1		
Uranomys foxi (56)	70.0	63.3	6.7	
Mus musculoides (28)	64.7	29.4		
Lemniscomys striatus (36)	70.6	76.5	17.6	
Praomys tullbergi (41)	100	44.4		
Lophuromys sikapusi (61)	20.0	78.3	85.0	10
Stochomys longicaudatus (3)	100	100		
Hybomys trivirgatus (11)	27.3	90.1	45.5	
Thamnomys rutilans (3)	100			

Temminck (1853), and the three-striped rat, *Hybomys trivirgatus* Temminck (1853), have body weight of between 50 and 100 g. Three species, Fox's brush-furred rat, *Uranomys foxi* Thomas (1912), Tullberg's rat, *Praomys (Rattus) tullbergi* Thomas (1894) and the shining thicket rat, *Thamnomys rutilans* Peters (1876) have body weight measurements of between 35 g to 50 g. The pigmy mouse, *Mus (Leggada) musculoides* Temminck (1853), has a body weight measurement of less than 10 g.

The stomach content weight varied between approximately 1 g in the pigmy mouse to a little over 3 g in some of the larger rats. The qualitative composition of the stomach contents in each species is shown in Table 2. The plant materials identified in stomachs included oil-palm fruit, maize grains, rice grains, cowpea pods and seeds, green leaves and roots. Insects found in stomachs included Dipteran larvae, ants, beetles, and variegated grasshoppers. Chewed polythene bags were found in the stomachs of two *R. rattus*.

Six species, *D. incomtus*, *M. natalensis*, *U. foxi*, *L. striatus*, *L. sikapusi* and *H. trivirgatus*, which live either in litter, or burrow in the soil all ate earthworms. *L. sikapusi* was found to eat earthworms more frequently than any other food items and it is the only rat that ate earthworm cocoons, myriapods and molluscs and could therefore be described as an invertebrate feeder. *H. trivirgatus* was also mainly insectivorous in its diet. Other rats and mice, besides *L. sikapusi* and *H. trivirgatus*, are mainly vegetable eaters and eat insects only occasionally and can, therefore, be regarded as potential pests of crop plants.

Records of pregnancies and the capture of juveniles were available mainly for March to October, during which period most species of rats and mice in the study were found to be breeding. The number of embryos per female varied from 3–5 in *R. rattus*. Pregnant female *R. rattus* were obtained in March and April and juveniles weighing between 13–70 g were trapped in all months from March to October indicating that breeding occured throughout these months. Only one pregnant *D. incomtus*, with 2 embryos, was captured in May but juveniles weighing from 30–52 g were captured in March, April, May and June indicating that reproduction of these species is also linked with the rains. The number of embryos per pregnant female in *M. natalensis* varied between 7–12. Pregnant females and also juveniles weighing 17–44 g, were captured in all months from March to September indicating that breeding occurred throughout this period. Pregnant *A. niloticus*, with 3–5 embryos per female were captured in April and May and juveniles were obtained in July, August and September indicating that breeding occurs in the rainy season. Pregnant female *U. foxi* with 2–6 embryos were captured in March, April, May and August. Juveniles of this species were also captured in March and in all months from July to October. Pregnant *L. striatus*, with 3–4 embryos were captured only in April and juveniles were obtained only in July and August. Two pregnant *M. musculoides* were captured in March (2 embryos) and April (1 embryo) and one juvenile in August. Pregnant female *L. sikapusi*, with 2–5 embryos were trapped in March, April, September and October and juveniles were captured in April, May, September and October. Three pregnant *H. trivirgatus* with 3, 4 and 5 embryos were captured in February and March and a juvenile weighing 41 g was captured in May. Two pregnant female *T. rutilans*, each with 2 embryos, were captured in April and December. One lactating *P. tullbergi* weighing 41 g with milk in its mammary glands was captured in April. No pregnant female or juvenile *Stochomys (Rattus) longicaudatus* Tullberg (1893) were obtained.

DISCUSSION

The names of rats and mice used in this report are those used by Rosevear (1969). There are a few species which were mentioned by this author for this region which were not captured during our investigations. It was not possible to determine monthly and local variations in the body weight of rats due to the small size of the sample.

It is suggested that to obtain maximum benefits from field investigations data should be collected in such a way that local, seasonal and sexual variations in body size can be determined. It is also desirable to use only the body weight of adults to compute the mean body weight for each species as the variations in the weight of juveniles are very large and may make results of different workers difficult to compare, more so when the proportion of juveniles vary in the different populations.

The determination of the diet of animals from stomach contents is not an easy task because in many cases the stomach contents are in an advanced stage of digestion before the analysis is made. Rats usually chew and reduce their food to very fine fragments before swallowing them and such deformed food materials may be difficult to identify even when they have not been changed by digestive juices. Our investigations, however, seek primarily to determine which species of rats are herbivorous feeders and therefore most likely to damage crop plants, for which our methods of analysis were very suitable.

The present report shows that two species, the rufous-bellied rat, *L. sikapusi*, and the three-striped rat, *H. trivirgatus*, are mainly insectivorous. *L. sikapusi* also feeds on oil-palm fruits and can cause considerable oil-palm fruit losses when it has a large population (Funmilayo, 1975). All the other rats and mice captured feed mainly on plant materials and these herbivorous species have been implicated in the damage to crop plants (Everard, 1966, 1968; Funmilayo, 1973).

We obtained pregnant females and juvenile *R. rattus*, a commensal species, in all months from March to October which confirms the Buxton's (1936) result — that *R. rattus* reproduces in all months of the year in Lagos. Similarly, we obtained pregnant females and juveniles of *M. natalensis*, a semi-domestic species, in all months from March to October. Pregnant female and juvenile *M. musculoides* were captured in March, April and August which agrees with Anada's (1974) results. Neal (1968) observed similar habits of breeding mainly during the raining period in four species, *L. striatus, M. natalensis, L. sikapusi* and *A. niloticus* in Uganda. Most of the species of rats and mice captured in the present study area were breeding, and therefore increasing in population numbers, within the period March to October when the rains fall annually and the bulk of food crops are planted. Rat damage to crop plants at this period is, no doubt, aggravated by this natural increase in the numerical abundance of those rat species that feed on crop plants.

Records of the reproduction of rats and mice, particularly the pest species, in Nigeria are incomplete and there is no doubt that a further study is needed.

Such a study should cover all months in a year and determine the seasonal and local variations in sex ratio, and the proportion of pregnant, lactating and non-breeding females in the populations of the different species. The litter size, the number of litters per female per annum and the survival rate from young to adults should also be measured. Only when we know these details can we forecast the trends in population numbers and successfully control those species that are responsible for the depredation of crop plants and stored products.

ACKNOWLEDGEMENTS

This work was carried out with funds and facilities provided by the Institute of Agricultural Research and Training, University of Ife, Ibadan for which we would like to thank the authorities.

REFERENCES

AMANDE, M. and FUNMILAYO, O. (1974) Some common diseases that can be transmitted from rodents to man. Research Bulletin No. 4. (Institute of Agricultural Research and Training, University of Ife, Nigeria.)

ANADU, P. A. (1974) The ecology and breeding biology of small rodents in the derived savannah zone of south-western Nigeria. (Ph.D. thesis, University of Ibadan, Nigeria.)

BUXTON, P. A. (1936) Breeding rates of domestic rats trapped in Lagos, Nigeria, and certain other countries. Journal of Animal Ecology, Vol. 5, pp. 53–65.

CHITTY, D. and SOUTHERN, H. N. (1954) Control of rats and mice. Vols. 1, 2, and 3. (Clarendon Press, Oxford.)

EVERARD, C. O. R. (1966) Reports on some aspects of rodent damage to maize in the Western State of Nigeria. (Research Division, Ministry of Agriculture and Natural Resources, Moor Plantation, Ibadan, Nigeria.)

EVERARD, C. O. R. (1968) A report on rodent and other vertebrate pest of cocoa in Western Nigeria. (Research Division Ministry of Agriculture and Natural Resources, Moor Plantation, Ibadan, Nigeria.)

FUNMILAYO, O. (1973) A general survey of the incidences and control methods of vertebrate pests of crop plants in the Western State of Nigeria. Research Bulletin No. 1. (Institute of Agriculture Research and Training, University of Ife, Nigeria.)

FUNMILAYO, O. (1975) The ecology and economic importance of the rufous-bellied rat, Lophuromys sikapusi sikapusi in Ibadan, Nigeria. Nigerian Journal of Plant Protein, Vol. 1, No. 1, pp. 17–22.

FUNMILAYO, O. and AKANDE, M. (1975) The black/grey rat, Rattus rattus rattus, as a pest in a rabbitry in Ibadan. Nigerian Journal of Forestry, Vol. 4, No. 1, pp 24–8.

MOSBY, H. S. (1963) Wildlife investigational techniques. 2nd Edition. (The Wildlife Society, Washington D.C.)

NEAL, B. R. (1968) Breeding seasons in Ugandan rodents. Journal of Animal Ecology, Vol. 37, pp. 7–8.

ROSEVEAR, D. R. (1953) Checklist and Atlas of Nigerian mammals. (Government Printers, Nigeria.)

ROSEVEAR, D. R. (1969) The rodents of West Africa. (British Museum [Nat. Hist.], London.)

17

SOME ASPECTS OF THE PHYSIOLOGY OF THE DOMESTICATED AFRICAN GIANT RAT

M. O. Olowo-Okorun

Department of Veterinary Anatomy and Physiology,
University of Ibadan, Nigeria

INTRODUCTION

African giant rats (*Cricetomys gambianus* Waterhouse) were captured from the bush and domesticated under standard conditions as described by Ajayi (1974). The domestication allowed the study of the influence of the confinement on the blood picture of the different generations of the male rats and the females were used for studies of the oestrus cycle. Both the male and female rats were used in a taste experiment.

MATERIALS AND METHODS

Twenty-four active looking adult rats, twelve of which were males and the other twelve females, were used for the experiments. The male rats were of four types, the wild freshly captured rats, and the first, second and third generations of the domesticated rats. There were three male rats from each of the four types. Blood was collected from the cut tip of the tails of the male rats directly into heparinized haematocrit tubes and also into heparinized containers. The packed cell volume was determined by the microhaematocrit method, and the red cell count, white cell count, and haemoglobin content were determined four times in each blood sample. The blood glucose level was determined in each blood sample using 4-aminophenazone as oxygen acceptor (Trinder, 1969).

The twelve adult female non-pregnant rats were isolated in twelve cages and vaginal smears were taken daily for fifteen consecutive days. The smears were fixed in 95 per cent alcohol and stained in Haematoxylin and Eosin (H and E) stain.

Two types of fluid were used for a taste experiment. One bottle contained 100 ml of water, while the other bottle contained a glucose solution (5 g of glucose made up to 100 ml with water). The two bottles had nozzles which could not be broken by the rats' teeth. The bottle containing glucose solution was introduced into the cage first and its nozzle was allowed to enter the rat's

142

mouth. The reactions of the rat to the solution were observed for three minutes before the nozzle was withdrawn from the rat's mouth. The reactions after the withdrawal were observed for two minutes before the bottle containing only water was introduced to the rat, and the same kind of observations were made. These procedures were repeated for four consecutive days using both the male and female rats.

RESULTS

The results for the blood variables are shown in Table 1. The blood of rats in the third generation of domestication had the highest values and the blood of the wild rats had the smallest for the blood glucose level, packed cell volume, red cell count and haemoglobin content. The differences in these values were significant for the packed cell volume, red cell count and haemoglobin content, $P<0.05$, while non-significant for the blood glucose level $P>0.10$. The results of white cell counts were the same for domesticated and wild rats.

For the oestrus cycle determination, the stained slides were examined under the microscope and Figures 1–5 represent sections from each of the different phases observed in five consecutive days of the fifteen days observation. Figure 1 shows a preponderance of leukocytes and few large nucleated cells. This stage represents the di-oestrus phase. Figure 2 contains mainly the large nucleated epithelial cells with no leucocytes and this represents the pro-oestrus phase.

TABLE 1

Blood values in the adult male African giant rats

	Wild rats (3)	First generation rats (3)	Second generation rats (3)	Third generation rats (3)
Mean packed cell volume in %	39.5 ±1.6	41.87 ±1.61	44.75 ±0.76	44.95 ±0.79
Mean red cell count in millions/mm³	4.194±0.17	4.950±0.21	5.525±0.23	5.725±0.16
Mean white cell count in thousand/mm³	5.521±0.19	6.054±0.22	5.782±0.24	5.956±0.18
Mean haemoglobin content in g/100 ml blood	10.82 ±0.61	11.55 ±0.68	12.75 ±0.72	12.94 ±0.71
Mean blood glucose level in mg/100 ml blood	80.1 ±2.4	81.4 ±3.1	82.6 ±3.4	83.6 ±3.9

Numbers in parenthesis indicate number of rats used.

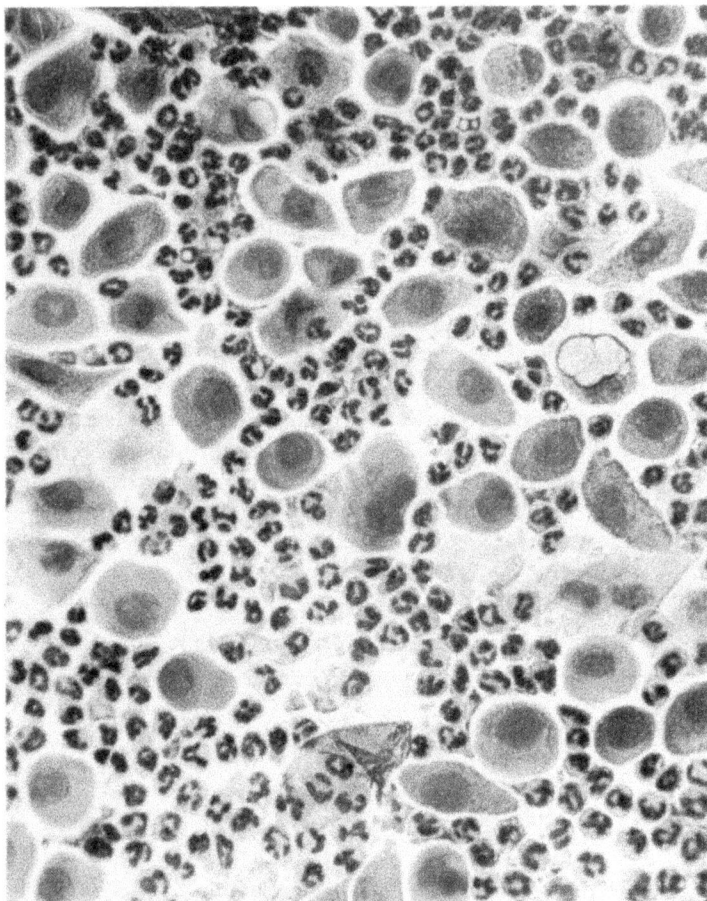

Fig. 1. Di-oestrous phase.

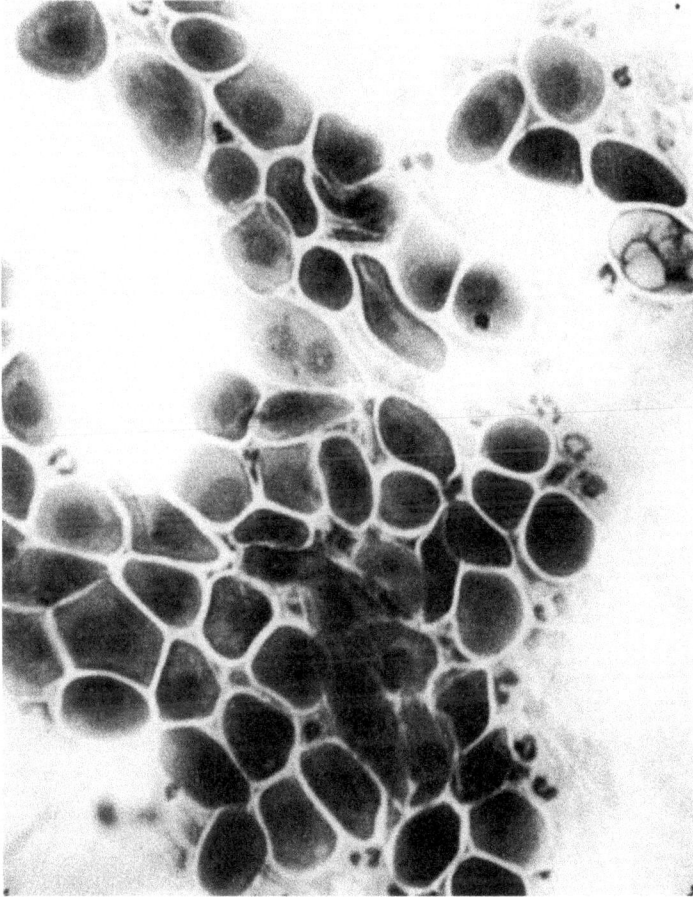

Fig. 2. Pro-oestrous phase.

Fig. 3. Oestrous phase.

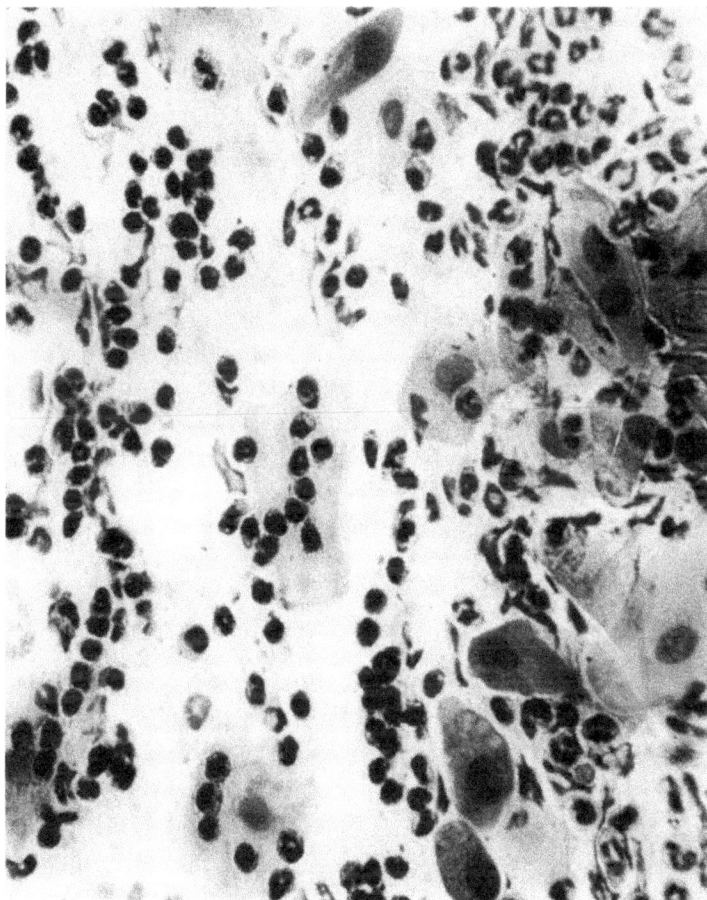

Fig. 4. Met-oestrous phase.

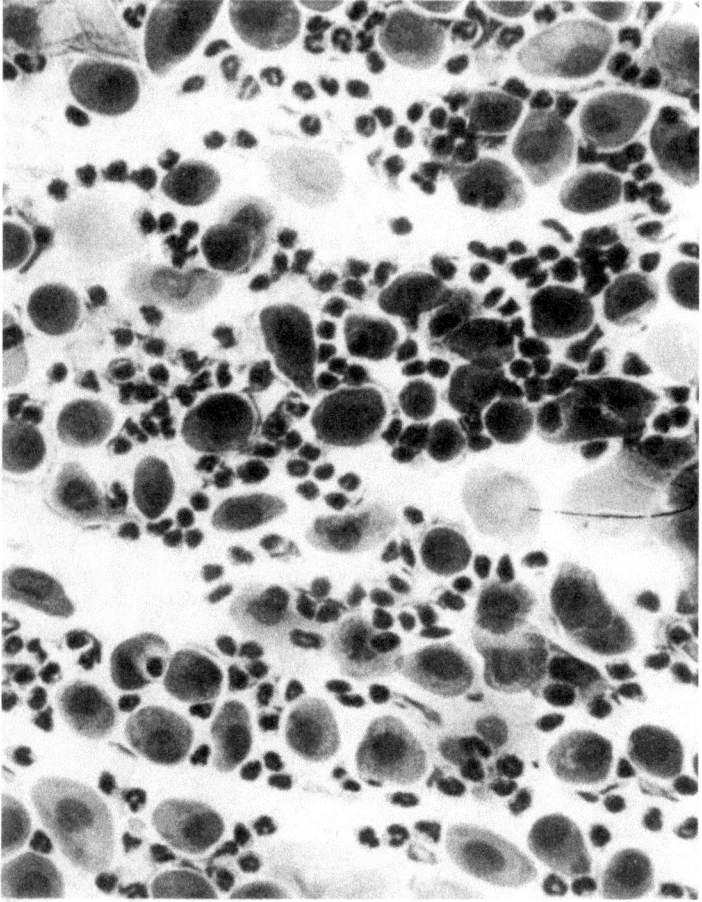

Fig. 5. Di-destrous phase.

Figure 3 shows large squamous cells, with small pyknotic nuclei and it represents the oestrus phase. In Figure 4, the leukocytes reappear and the remnants of the large squamous cells with pyknotic nuclei of Figure 3 are seen. This phase represents the met-oestrus phase. In Figure 5, similar cellular structures found in Figure 1 are seen, and this also represents the di-oestrus phase. The results from the stained vaginal smears show that the oestrus cycle lasts for four to five days.

The rat's reactions to the two types of fluid were different. The rats were observed to be restless when the glucose solution dropped into their mouths, clung to the nozzle of the bottle and attempted to pull the bottle away. When the glucose solution was withdrawn, the rats moved round the cage as if in search of the nozzle. When the bottle containing only water was introduced, the rat grabbed the nozzle and quickly released its grab when the water dropped into the mouth. The anxiety which accompanied the introduction of the glucose solution was lost and the rats were no longer struggling to pull the nozzle away from the cage. When the water was first introduced the rat showed some excitement, clung to the nozzle but these reactions were greatly magnified when the glucose solution replaced the water.

DISCUSSION

The changes observed in the blood values of the domesticated rats could be due to the overall changes in the environment to which the rodents were exposed. They were given better and regular diets in the cages, and they were adequately protected from the stress and the dangers to which the wild rats were constantly exposed. The advantage of this to rat breeding as a human food resource would be that improved blood condition will enhance better performance of the domesticated animals and when these animals are used for breeding, better offspring would result.

The duration of the oestrus cycle (4–5 days) is important for rat breeding exercises. The male rat could be introduced on the day of oestrus, when the female rat would give maximum co-operation until after copulation. Also, the average gestation period could be carefully worked out. Further work on reproduction could also stem from the oestrus cycle determination.

The rats have been exposed only to glucose solution and water and the observations have led to the possible belief that the African giant rats could differentiate between the two types of fluids. Further work is going on with other types of taste tests and later the sense of smell of these rats will be tested.

ACKNOWLEDGEMENTS

The author is grateful to Dr S. S. Ajayi for allowing the use of his domesticated rats. Mr T. A. Asojo assisted in the staining of the vaginal smears. Miss A. O.

Makinde and Mr Omolaja Mummuney and Mr F. O. Ogunji offered technical assistance. The fund for the work was from a Senate Research Grant, University of Ibadan.

REFERENCES

AJAYI, S. S. (1974) Preliminary observations on the biology and domestication of the African giant rat (*Cricetomys gambianus* Waterhouse) in Nigeria. *Mammalia,* No. 3.
TRINDER, P. (1969) Determination of blood glucose using 4-aminophenazone as oxygen acceptor. *Journal of Clinical Pathology,* Vol. 22, p. 246.

SECTION 3
FIRE

18

THE IMPLICATIONS OF WOODLAND
BURNING FOR WILDLIFE MANAGEMENT

W. A. Rogers

Department of Zoology,
The University of Dar es Salaam, Tanzania

INTRODUCTION

A considerable literature exists on the effects of fire in woodland and grassland environments in Africa. Major reviews have been published by Daubernmire (1968), West (1965), Hopkins (1965) and others. These publications have not however, removed the controversy surrounding the use of fire; 'to burn or not to burn, to burn early or late in the dry season, to burn annually or at what intervals?' With such a choice of fire regimes, variety of habitats, soil types, and climatic regimes in Africa; and the differing end point requirements of foresters, agriculturalists, livestock managers and wildlife managers, it is not surprising that controversy exists.

In addition, there is often a considerable difference between the ecological optimum fire regime, and what can be practically accomplished with only limited finance, staff and equipment. Furthermore, land use practice may require differing fire regimes from the ecological optimum. All too often fire management must be a compromise between the ecological needs, land use needs and practical limitations.

THE RANGE OF FIRE EFFECTS

Fire can affect the following environmental variables either directly or indirectly:

(1) Soils by:
- (a) affecting numbers and rate of activity of soil organisms,
- (b) removing or changing rates of soils organic matter formation and accumulation,
- (c) affecting surface compactness,
- (d) affecting amounts and availability of essential nutrients,
- (e) affecting soil water retention properties,
- (f) removing soil surface horizons through surface run off and sheet erosion.

151

  (2) Land surface by:
       (a) affecting degree and rates of surface erosion by effects on soil and
            vegetation cover,
       (b) effects on water movement.
  (3) Water by:
       (a) changing rates of transpiration and evaporation,
       (b) changing rates of permeability and subsurface flow,
       (c) affecting amounts and rate of sedimentation,
       (d) changing stream and river structure, through bank and surrounding
            vegetation destruction.
  (4) Vegetation, both directly and indirectly through the habitat effects
       mentioned above, e.g.
       (a) changing direction and speed of vegetation succession,
       (b) affecting plant biomass, structure and shape,
       (c) affecting plant phenology,
       (d) affecting plant quality in terms of nutrient content and availability.
  (5) Animals by:
       (a) changing the shape or amount of cover,
       (b) changes in plant palatability and availability,
       (c) indirectly altering water availability,
       (d) causing death or injury (especially lower orders of animals).
It is obvious that many of these effects are related and interacting, thus the
study of fire ecology and implementing its management is extremely complex.

TYPES OF FIRE REGIME

A cool fire is an early dry season fire, set before the grasses have completely
dried out. It moves close to the ground, shooting up to grass tops as they are
encountered. Temperatures rarely reach 300°C and are minimal at 2 cm below
ground level. Tree tops escape damage as do the denser shrubs and greener
'shade loving' grasses. Dead wood is only slowly consumed. Small tracks, water
courses, valleys and ridges can act as barriers to these fires, and a heavy dew fall
can extinguish them. As a result they rarely cover very large areas.
  A hot fire is set at the end of the dry season, when the grass cover is
completely dry. Fires move rapidly at 1–2 m above ground, temperatures can
reach 600°C or more, and temperature effects can reach down to 5 cm below.
Tree tops are scorched and leaves killed, shrubs and seedlings are engulfed.
Dead wood is rapidly consumed. Small barriers (see above) can be jumped and
such fires can cover large areas in a short space of time.
  Cool and hot fires are not necessarily determined by calendar months; the
onset and termination of rains, plant species and stage of maturity of grass and
its water content, can all affect fire temperatures. In south-west Tanzania,
June–July fires may be termed cool fires and October fires hot. Zambia can be

up to a month earlier. Wind direction can affect fire temperatures; a back fire moving slowly with its heart near the ground can subject plants to higher temperatures for a longer time than head fires moving rapidly a metre or so above the ground.

The frequency of fires will have a major influence, for example a gap of several years with no fire will allow the later burning of larger amounts of material with a hotter and more destructive fire.

A FIRE REGIME FOR WILDLIFE MANAGEMENT

A land use policy statement is essential before a fire regime can be chosen. Softwood foresters, for example, would wish to exclude all fires — while agriculturalists during land clearing operations may favour late fires for a number of years.

For wildlife management a major objective is the provision of suitable habitat for the game resource. The bulk of our African wildlife in terms of species and numbers of individuals are grazers, consequently the manager must aim to provide a sufficient grass cover of adequate nutrient value. As conservationists we are also interested in the continuation of other less noticeable species, such as forest duikers, requiring different more specialized habitats. Consequently there is a need for the maintenance of habitat variety. Resource use may provide difficulties, for example, a non-burnt 2 m high grass cover is not conducive to game viewing or hunting and these considerations of land use may affect the fire policy.

For most practical purposes there are two major fire effects to be considered. One is important in the long term, dealing with vegetation cover, its succession and change. The other is of annual importance, the structure of the grass cover, its biomass, availability and palatability.

Much evidence can be taken from a consideration of the history of fire, as man has been using fire for at least 53 000 years in Africa (Phillips, 1965), and natural fires have long been a feature of the environment (West, 1965). Most authorities agree that the present 'miombo' or similar 'guinea savannah' woodlands of East, Central and West Africa have been derived from an earlier thicket or forest vegetation type since the last major pluvial period some 20 000 B.P. This forest destruction has been caused by the effects of a drier climate, widespread fire and felling by agricultural man.

One can see many adaptations to a fire regime within the woodland habitat. Trees frequently have thick corky or flaking bark, rapidly forming scar tissue, dry season deciduousness, thick woody seeds, and some are even dependent on fire for germination, e.g. *Pterocarpus angolensis*. The development of the suffruticose and die-back habits in shrubs and seedlings are further fire adaptations. Animal adaptations can also be seem, e.g. the fire camouflage of plover eggs, and the seasonality of mammal calving.

THE EFFECTS OF DIFFERING FIRE REGIMES

Here, examples are drawn from S.E. Tanzania — the short-term burning experiments mainly dealing with ground cover changes at the Miombo Research Centre, Selous Game Reserve, Tanzania (8°30′E) and the long-term (40 year) experimental fire plots of the Zambian Forest Department, Ndola are discussed, (Trapnell, 1959).

No burning

This is favoured by foresters and intensive agriculturalists who can control shrub growth by other means. In most woodland habitats, no burning leads to thicket growth and so for the management of grazing herbivores it is not recommended.

The Miombo woodland complex and the associated *Combretum* and open woodland types are a fire held sub-climax vegetation type. Stopping fire will allow a reversion to the climax type, a form of forest or thicket depending on climate and soils.

The first effect of no fire is the retention of the previous year's coarse fibrous grass growth and a possible delay in the rapid production of new green growth. Seedlings and shrubs are not burnt back and growth continues unchecked. More seedlings and herbs can germinate. Litter accumulates on the ground surface, shading growing shoots and the soil surface.

An invasion by fire-intolerant species often follows. In many areas of the Selous, species of *Landolphia* and *Vernonia* are often dominant amongst the invaders. Both of these plants are straggling shrubs which clamber over grasses, seedlings and small trees forming dense tangles, which excludes light and hence the ground layer vegetation. Within three or four years there is a tangled mat of herbaceous vegetation, both herbs and grasses, and a dense ground litter. Through this are innumerable seedlings and shrubs which have been allowed to mature.

The fire-climax coarse grass species, e.g. *Hyparrhenia*, *Themeda*, *Loudetia* and *Andropogon* are checked and begin to disappear. The build up of litter and presence of old stems as well as increased shade are believed to cause their decline. Eventually a stage is reached when the seeds of fire-climax trees cannot germinate, seedlings and saplings, which were already present, are smothered and disappear.

The next stage is one of colonization by thicket species: *Commiphora*, *Croton*, and *Uvaria* being early arrivals in many cases, along with thicket grasses, such as *Leptochloa* sp. The end point which will take many years (possibly over 100) is a thicket or forest community, the composition of which depends on soil, site and climatic factors and the distance to other communities as a seed source.

Late burning

This is a fire regime favoured by many agricultural land managers which is used in land clearing for rural or industrial development.

When a late fire passes through an area all the ground cover is consumed — herbs, grasses and small shrubs. Dry leaf-canopies can flare and wet leaf-canopies are scorched and shrivelled. Tree boles can be burnt leaving fire scars — sources of infection and sites of deep scar tissue, which render the tree almost useless for valuable timber production. All leaf litter is engulfed and dead wood consumed, the area is laid completely bare. Many of Africa's woodlands show a leaf flush in the late dry season, before the first rains. If fires are set after the flush the effects will be even more devastating on the woody cover. There is little in the way of woody vegetation that can survive this treatment year after year. Seedlings and small shrubs which have not developed deep rooted shrubs will be cut back to ground level and produce new shoots the next year. Repeated late burning will, however, slowly kill these shrubs as well.

The coarse fire-climax grasses are favoured by this treatment and continue to produce dense growth year after year, the basal area of these species may increase but the total ground layer basal area made up of coarse bunch grasses and the smaller grasses, sedges and herbs will decrease due to the disappearance of the last three classes. The majority of fire-intolerant or semi-tolerant tree species will be killed. New trees of fire-tolerant species will be unable to germinate or mature. Gradually an open, wooded, tall grass land will be formed with coarse grasses showing vigorous growth, a few coppicing thick leaved shrubs, and a few fire-tolerant trees.

Early burning

Early burning is favoured by many wildlife and wilderness managers, agriculturalists and by those who, rather favouring no burning, but will accept an early burn as the only practical alternative to preventing late burns.

The effects of early burning are a combination or a mid-point between the two regimes discussed above. It tends to preserve the *status quo*. Early burning will remove old coarse grass and stimulate fresh new shoots. It allows more fire-sensitive grasses to continue if not to flourish. Some seedlings and shrubs are killed, the majority merely cut back to varying heights depending on the position of the growing buds. Anthill thickets continue, although they do not spread. A woodland formation of dense to sparse grass cover with a definite ground and shrub layer, and woody canopy remains.

Combinations of the three regimes produce mid-way effects, although a fierce, late fire can show effects on the environment for several years. A gap in burning for a few years followed by a late fire allows a far greater amount of combustible material to accumulate resulting in much fiercer fires. A few years gap in burning will manifest itself in a different shrub size or age structure in an otherwise uniform community.

FIRE AND GRASS PRODUCTION

Much of the Selous Game Reserve shows a rapid flush of perennial grass and tree leaf following early and mid-season fires. Exceptions are ridge tops where the water-table is low, and alkaline clays where soil moisture appears to be unavailable for plant growth. This flush is a common phenomenon throughout Africa and is deliberately induced by pastoralists, hunters and others.

Tall-grass sward in the woodlands of the Selous offers very poor dry-season grazing to herbivorous animals. Grasses are tall, up to 2 m high, heavily lignified, little or no fresh green growth and a leaf crude-protein content of some 1.8 to 4 per cent. Biomass may be high — up to 500 g dry matter per m^2 but nutrient content and palatability is low.

An early season fire will remove the coarse, dry grass and stimulate a flush of green growth which, although of low biomass, perhaps 10 g dry matter per m^2 can have a crude-protein content of 12 to 20 per cent and it is readily available and palatable to grazing animals.

The cause of the flush is still unclear, but has been attributed to temperature stimulation or the effect of massive fertilization following ash deposits and dew fall becoming available to the plants. It is likely that the real cause is the removal of old and dead plant material releasing the plant from a heavy water requirement for support of old tissues and also to the removal of pressure from the apical meristems of new shoot tips.

Perennial grasses translocate food reserves following flowering in the late rains and in the early dry-season. These reserves are used either in the flush induced by fire or, with no fire, in normal growth with the first rain. An early fire stimulates use of this food reserve, a late fire after the first rain showers will destroy such growth leaving the plant with no reserves for further production. A very early flush with no dry season showers will eventually wilt but the eventual rains cause growth in these same shoots.

We are fortunate in the Selous that early fires travel relatively short distances, and so burning takes place in a patchwork or mosaic fashion, creating a fresh flush of growth in several areas throughout the dry season.

Animal calving periods appear to be correlated with the high nutritive content of the dry season flush. Several animals, notably hartebeest, sable, impala, eland, wart-hog and zebra, calve in the dry season. Wildebeest in East Selous calve in late November when the annual grasses grow in the first rains, but some 80 km to the west across a major forest barrier, where there is no annual sward, wildebeest calve in early October coinciding with the dry-season perennial grass flush.

A FIRE POLICY FOR THE SELOUS GAME RESERVE

The Selous Game Reserve of some 40 000 km² encompasses 5 major vegetation types:
(1) Woodland communities of varying species composition and density.
(2) Forest thicket communities on sandy soils.
(3) Thicket communities on alkaline clay sands.
(4) Wooded short grassland communities.
(5) Grassland, edaphic flood-plains and swamps.

Woodlands

The policy here is to burn as early as possible in the dry season, and to attempt to achieve a patchwork, or mosaic, of burns from July to September, avoiding the more devastating late burns of October–November. Aerial survey shows that up to 60 per cent of the woodland community burns by late September each year. Fires are set early to create high quality, dry-season grazing and to avoid shrub invasion of the woodland ground layer.

Forests and thickets

The forests and thickets on sandy soils are associated with major water-shed ridges. They are extensive in area and from an analysis of aerial photographs from 1948 to 1965 they are decreasing only very slowly or not at all. These communities provide browse to elephant and the complete habitat for red and blue duiker and suni. They are not used by grazing animals. More importantly they are of value in water conservation, in that every major valley from a forested water-shed bears permanent springs and seepage lines. As such, the forests must be protected. Selous policy is not to open up major forest areas by roads or tracks. Fire-breaks are not practicable with present financial resources. Succesive years of late, hot fires would encroach on forest edges but early cool fires appear to have little permanent effect. Aerial photographs show the remains of old villages and cultivation in the forests (pre 1947). Where these are enclosed, succession back to closed thicket is virtually complete. Riverine forests are of similar importance but much more fragile and in places are disappearing due to fire encroachment from the outside and river widening from the inside. Degradation of these riverine forests is perhaps the most important fire-associated problem in the Selous.

Scattered tree short grassland

Grasses here are predominantly broad leaf sweet-veld, many of them annuals. With the exception of *Panicum infestum* none flush after burning on the alkaline, hard pan soils. The area carries high game-densities in the rainy season (75 animals per km²) which gradually move into the longer grass woodlands when they burn and flush. The annual grasses gradually decompose and break-down due to death and trampling and fires can be difficult to set due to lack of material grasses. Burning in this area is of no ecological value except an occasional late burn to prevent extensive growth of *Acacia* and *Combretum* species, which could suppress grass production. In fact the unburnt annual sward does provide some forage for grazing animals throughout the dry season.

Edaphic grasslands

Flood plain grasslands in the Selous provide a late dry season refuge for zebra and wildebeest. Burnt and regenerating areas carry high numbers of animals whereas the unburnt 2 m high grassland remains unused except by reedbuck and buffalo. As such these areas are burnt in the middle dry season (the earliest practical time) to afford high quality grazing.

DISCUSSION

This brief paper makes no attempt to discuss all aspects of fire-induced effects, changes and problems. It does hope however to focus attention on some of the sources of controversy in the present literature. The wildlife manager faced with the problem of 'to burn or not to burn' must start off with a clearly defined objective and policy for his wildlife estate. The manager can gain valuable clues from the history of burning in the area and from examining past aerial photographs. Has a vegetation change taken place in the past, if so is change proceeding in the direction the manager desires? If the estate is in an area of potential thicket or forest-climax vegetation then management for grazing animals must include burning.

In areas where dry-season flushing does not occur due to low rainfall, unsuitable soil types or predominantly annual grasses then early burning is not beneficial and a better policy may be one of infrequent late burns.

In some grassland areas the high density of large animals such as hippopotamus, elephant or buffalo can take the place of fire by grazing and trampling the tall grasses, e.g. lake Rukwa in South-West Tanzania (Vesey-Fitzgerald, 1965). In this case burning would be superfluous and could even lead to a decrease in numbers of large animals.

It may be that some other management problem is more pressing than that of stimulating dry season grazing. For example, it could be argued that to reverse the present trend of woodland destruction in Lake Manyara National Park (Northern Tanzania) it is of paramount importance that fire must be excluded.

As has been stressed in this paper fire management must often be a compromise between conflicting viewpoints. Only when a definite land-use policy and objective has been decided can one plan a satisfactory fire management policy.

ACKNOWLEDGEMENTS

This paper was prepared when the author was a member of the Miombo Research Centre. I am indebted to the staff of the Centre and to the Director of Game in Tanzania for facilities to undertake this work. During this period the Centre was funded by a generous grant from the Government of Denmark.

REFERENCES

DAUBENMIRE, R. (1968) Ecology of fire in grassland. *Advances in Ecological Research*, Vol. 5, p. 209.

HOPKINS, B. (1965) Observations on savannah burning in the Olokemiji Forest Reserve, Nigeria. *Journal of Applied Ecology*, Vol. 2, p. 367.

PHILLIPS, J. (1965) Fire as a master and servant: Its influence on the bioclimatic regions of trans-Sahara Africa. *Tall Timbers Fire Ecol. Conf.* Vol. 5, p. 7.

TRAPNELL, C. G. (1959) Ecological results of woodland burning in Northern Rhodesia. *Journal of Ecology*, Vol. 47, p. 129.

VESEY-FITZGERALD, D. F. (1965) The utilisation of natural pastures by wild animals in the Rukwa Valley, Tanganyika. *East African Wildlife Journal*, Vol. 3, p. 38.

WEST, O. (1965) Fire in vegetation and its use in pasture and subtropical Africa. Commonw. Bur. Past and Crops, Farnham Royal, Bucks, UK. P. 53.

APPENDIX

A list of animal names mentioned in the text:

Buffalo — *Syncerus caffer* (Sparrman, 1779).
Blue duiker — *Cephalophus monticola* (Thunberg, 1789).
Red duiker — *Cephalophus natalensis A.* (Smith, 1834).
Eland — *Taurotragus oryx* (Pallas, 1776).
Hartebeest — *Alcelaphus lichenststeini* (Peters, 1849).
Reed-buck — *Redunca redunca* (Pallas, 1777).
Impala — *Aepyceros melampus* (Lichtenstein, 1812).
Sable — *Hippotragus niger* (Harris, 1838).
Suni — *Neotragus moschatus* (On Dueben, 1846).
Wildebeest — *Connochaetes taurinus* (Burchell, 1823).
Zebra — *Equus burchelli* (Gray, 1824).

19

AN APPRAISAL OF THE SAVANNAH BURNING PROBLEM

J. K. Egunjobi

Department of Agricultural Biology,
University of Ibadan,
Nigeria

INTRODUCTION

There is a long-standing controversy about savannah burning. This controversy arises out of the conflicting interests of foresters and soil conservationists on the one hand and the grazier on the other. Successive governments in Nigeria, and elsewhere in Africa have at one time or another shown concern about savannah burning through legislation. In 1937, the Forestry Ordinance Law was enacted to prohibit the lighting of fires in forest reserves in northern Nigeria. As this was only limited to protected areas, the Burning of Bush Order was enacted in 1940, which empowered native authorities to fix the season at which bush fires may be lit and to penalize defaulters. As late as 1970, a similar law was enacted by the Military Governor of the Western State in view of the prevailing drought at that time. Similar laws controlling savannah burning have been promulgated in other parts of Africa. (See for example the Uganda Careless Use of Fire Prevention Ordinance 1920, the Sierra Leone Bush Fire Prevention Ordinance 1924, the Nyasaland Bush Fire Ordinance 1932, the Northern Rhodesia Bush Fire and Trespass Ordinance 1948, etc.) Despite all these government actions savannah burning is still practised. This paper discusses the reasons for which savannah is burnt, the conflicts of interests arising from savannah burning, the effects of burning on the ecosystem, and the possible use of fire in the management of range-lands for wildlife.

REASONS FOR WHICH SAVANNAH IS BURNT

The various causes of savannah fires has been discussed in some detail by West (1965). Fires may result from a natural phenomenon such as lightning, or through accidental action by man, e.g. dropping a cigarette butt, or as a result of deliberate action of man. A high proportion of savannah fires are started by man. The main reasons for which man burns savannah annually or periodically are:

(1) To prepare land for cultivation.

(2) To flush out animals during hunting expeditions.

(3) To remove old unpalatable growth to promote new grass flush.

(4) To control brush and encroachment of rangeland by woody species.

(5) To destroy parasites, such as ticks, which carry and transmit stock diseases. For example it has been suggested (Morris, 1944) that bush burning may be used in the control of Tsetse flies (*Glossina* spp.) in northern Nigeria.

In Nigeria, most annual fires are caused deliberately during land preparation and by graziers and hunters.

CONTROVERSY ABOUT SAVANNAH BURNING

The interests of foresters and graziers are parallel and cannot be reconciled. The foresters main interest is to encourage the regeneration of trees and protect those existing, while graziers are mainly interested in promoting the growth of grasses and the herbaceous flora. Foresters therefore like to prevent fire while graziers encourage it. The conflicting interests of foresters and graziers have been reflected in legislation and official recommendations on burning. In countries where animal production is paramount then late burning is encouraged, but where forest growth is considered most important as in Nigeria and other West African countries, emphasis is on early burning, because this is not hot enough to destroy the trees.

The need to protect soil from erosion has added to the controversy on burning. Obviously a good vegetable cover and mulch protects the soil against the kinetic force of rain drops and reduces the likelihood of erosion. This has led Dundas (1944) to say that 'In a tropical country like Nigeria where the rains are torrential, the winds frequently strong and the sun very hot, even a slight disturbance of the vegetal mantle may have very considerable ill effects'. He observed that the most important results of undue interference with the vegetation cover are:

(1) Degradation of soil through depletion of organic material and decimation of micro-organisms through insolation.

(2) Reduced availability of rainfall through increased run off and lessened percolation.

(3) Seasonal disappearance of formerly perennial streams.

(4) Water and wind erosion.

(5) Flooding and silting up of low lying area.

He went further to say that 'the extensive evils of bush fires are coming to be fully appreciated in Nigeria, and it is realised that they are resulting in a general deterioration of the sylvan conditions, the maintenance of which is essential for the upkeep of the conditions of life and standard of living of an agricultural community in the tropics'.

These statements were written 1944 and James Dundas was a forester. How true are they? A considerable amount of research on the role of fires in savannah ecosystems has been conducted since Dundas expressed his views. Empirical data indicates that his statements need to be modified.

It may be generalized that in the tall humid savannah burning *per se* does not lead to erosion. As shown in a previous paper (Egunjobi, 1971), there was no evidence of erosion when savannah dominated by *Andropogon gayanus* was burnt during two consecutive years at Fasola near Oyo (western Nigeria). Nye and Greenland (1959) had earlier reached a similar conclusion when they recorded that 'the stable land forms developed in the subhumid conditions are little subjected to erosion even though the protective ground cover is periodically removed by fire'.

The fact of the case is that the burning is done when there is practically no rain. Furthermore, new growth of grass starts 5 to 6 days after burning, so that the soil is not left bare for too long. By the time the heavy rains begin, the ground has become completely covered by vegetation. However, when burning is followed by ploughing or immediate hard grazing, erosion may set in.

The suggestion that burning also leads to a loss of soil organic matter is not justified by empirical data. Data provided by Moore (1960) indicated that soil from savannah burnt early in December for 30 years contained more organic matter (4.3 per cent) than soils from areas protected for fire (3.8 per cent). However, soils from savannah burnt late (March) for the same length of time contained less organic matter (3 per cent). Similarly the author did not find any immediate reduction in organic matter content following burning in December, January and February.

EFFECTS OF BURNING ON THE ECOSYSTEM

Burning has very drastic effects on the savannah ecosystem. The degree of this impact is dependent on:

(1) The total amount of combustible material present.
(2) The dryness of the material.
(3) The season and time of burning.
(4) The prevailing weather conditions at the time of burning.

A comprehensive review of the effect of burning has been made by Daubenmire (1968). The effects of annual or periodic fires may be short term or long term. The short-term effects centre around the destruction of most of the vegetal cover and the sudden release of ash elements. These ash elements are soon utilized by plants or leached by rain. Some of the short-term effects have been fully discussed in a previous paper (Egunjobi, 1971).

The most obvious of the long-term effects is on the species composition of the vegetation. This is perhaps the most noticeable effect of burning when adjacent protected areas are compared. Comparisons of this type have been published by Charter and Keay (1960), Onochie (1961), Ramsay and Innes (1963) and Hopkins (1965). Generally areas protected from fire have become more wooded with increases in fire tender species, while areas subjected to burning have fewer trees and retain a denser grassy flora.

EFFECT OF ANNUAL BURNING ON THE WILDLIFE

One major impact of annual or periodic fires is on the animal life. There are no published accounts of the effects of fires on wildlife for the West African region but a few studies have been reported for central and southern Africa. See for example Lemon (1968) referring to Malawi.

Burning of vegetation has the following adverse effects on the wildlife populations:

(1) Destruction of habitat.
(2) Removal of food resources — mainly for herbivores.
(3) Destruction of young and eggs.
(4) Exposure to predators.

On the credit side the main beneficial effects of burning are:

(a) Removal of old dead grasses which are unpalatable and which reduce current production and hinder free movement of animals.
(b) Increase in primary production (see Egunjobi, 1973).
(c) Higher nutritive value of new swards of grass from the burnt areas. In a study conducted at Fasola near Oyo in Nigeria, Egunjobi (1973) recorded a higher crude protein content (8 per cent) for new grass growth from burnt areas than in the unburnt area (4 per cent). However this advantage was soon lost as the plants matured. A similar observation has been reported by Lemon (1968) in Malawi.

There are no quantitative accounts of animals dying as a result of fires. Most ungulates, rodents and reptiles will escape from the fire front, and some times animals escaping with burnt hair have been noticed. Occasionally carcases of animals have been found following a burn but it is not known whether these have been killed by fire or by predators.

Obviously, one will not expect young or ailing animals to escape the holocaust. Similarly eggs of ground nesting birds will be destroyed and occasional charred eggs of nesting birds have been observed at the Fasola burning experiment.

It may be assumed that burrowing animals are able to escape by running into their burrows and riverine vegetation which is usually not as dry as the surrounding vegetation. This riverine flora is of considerable importance in providing sanctuary for animals escaping from burnt areas.

FIRE IN RELATION TO WILDLIFE MANAGEMENT

The beneficial effects of periodic burning of savannah have been listed in the previous section. Notable among these is the higher nutritive value of herbage from burnt areas. In any range-land or savannah where grazing is the prime objective, there is probably no alternative to periodic burning. However, this has to be done in such a way as will not affect the ecosystem adversely. According to Miller and Watson (1973) the distribution of animals in space and time is often related to the quantity and quality of food available. A spatial distribution of good quality forage may be achieved through habitat manipulation by prescribed burning. At the Fashola burning experiment, cattle were found to be attracted to recently burned areas. Such attractions to areas burnt have also been reported for wild mammals by Lemon (1968).

Although it is generally accepted that wild animals exploit the range better than domesticated livestock very little is known about their food preferences. However, it must be noted that most animals prefer the young buds of plants and new sward of grasses provided on a recent burn to older parts of the same plants.

Besides providing luscious food for the animals, prescribed burning can be used in maintaining a desirable plant species on the range. However in the manipulation of the habitat through prescribed burning, one should take cognizance of the animal population in the area — a large population grazing a small area may result in over-grazing with subsequent deterioration of the range.

REFERENCES

CHARTER, J. R. and KEAY, R. W. J. (1960) Assessment of the Olokemeji fire control experiment (investigation 254) 28 years after institution. *Nigerian Forest Information Bulletin,* Vol. 3.

DAUBENMIRE, R. (1968) Ecology of fire in grasslands. *Advances in Ecological Research.*

DUNDAS, J. (1944) The burning question. *Farm and Forest*, Vol. 5, pp. 8–10.

EGUNJOBI, J. K. (1971) Savannah burning, soil fertility and herbage productivity in the Derived Savannah Zone of Nigeria. *Proceedings of the Symposium of Wildlife Conservation in West Africa.* (I.U.C.N. Publication.) Vol. 22, pp. 52–8.

EGUNJOBI, J. K. (1973) Studies on the primary productivity of a regularly burnt tropical savannah. *Annals of the University of Abidjan, Series E,* Vol. 6, pp. 157–69.

HOPKINS, B. (1965) Observations on savannah burning in the Olokemeji Forest Reserve, Nigeria. *Journal of Applied Ecology*, Vol. 2, pp. 367–81.

LEMON, P. C. (1968) Fire and wildlife grazing on an African plateau. *Proc. Tall Timbers Fire Ecology Conf.*, Vol. 8, pp. 71–87.

MILLER, G. R. and WATSON, A. (1973) Some effects of fire on vertebrate herbivores in Scottish highlands. *Proc. Tall Timbers Fire Ecology Conf.,* Vol. 13, pp. 39–64.

MOORE, A. W. (1960) The influence of annual burning on a soil in the Derived Savannah Zone of Nigeria. *7th Trans. International Cong. Soil Sci.*, Vol. 4, 257–64.

MORRIS, K. R. S. (1944) A large scale experiment in the eradication of tsetse. *Farm and Forest*, Vol. 5, pp. 149–56.

NYE, R. H. and GREENLAND, D. J. (1959) The soil under shifting cultivation. *Tech. Commun. Commonw. Bur. Soils*, Vol. 51, p. 156.

ONOCHIE, C. F. A. (1961) A report on the fire-control experiment in Anara Forest Reserve (Investigation 222). *Fed. Dept. For. Res. Tech. Note*, Vol. 14.

RAMSAY, J. M. and INNES, R. R. (1963) Some quantitative observations on the effects of fire on the Guinea Savannah vegetation of Northern Ghana over a period of 11 years. *African Soil*, Vol. 8, pp. 41–85.

WEST, O. (1965) *Fire in vegetation and its use in pasture management with special reference to tropical and sub-tropical Africa.* (Mimeogrd. Publ. Commonw. Bur. Past. Fld. Crops.)

20

EFFECTS OF BURNING TREATMENTS ON THE STANDING CROP AND LITTER DEPOSIT IN THE GRASSLAND SAVANNAH OF THE KAINJI LAKE NATIONAL PARK

T. A. Afolayan

Department of Forest Resources Management,
University of Ibadan, Nigeria

INTRODUCTION

The effects of fire and grazing on the structure and productivity of grassland have been investigated by several ecologists and range managers. Rodgers (1970) gave an account of the effects of fire on Miombo Woodland in Tanzania. This account is of considerable interest to ecologists and range managers working in the West African savannah woodland because of the similarity to Miombo.

Rodgers' paper is a very comprehensive one including the effects of different fire regimes on trees, shrubs, grasses, soils, water-sheds, dry forests and woodlands. Hopkins (1962, 1963, 1964, 1965) working at Olokemeji Forest Reserve in Nigeria made several observations on the effects of fire on tree and shrub populations, soils and herbs. Similar experiments have been carried out by Moore (1960) in the derived savannah zone of Nigeria, Onochie (1964) in the Sudan zone of Nigeria, Ramsay *et al.* (1963) in the Guinea savannah vegetation of Northern Ghana and Rains (1963) at Samaru in Northern Nigeria. Egunjobi (1971, 1973) carried out a series of experiments at Fasola ranch in the derived savannah of Nigeria on savannah burning, herbage productivity, soil fertility, litter deposit, and organic and mineral fluxes in the savannah. Either by design or accident none of the above experiments had been carried out in a typical wildlife habitat. All efforts so far have been concentrated on cattle ranches or in forest reserves with little game. The experiment described in this paper is the first of its type in a wildlife habitat in Nigeria.

This experiment was set up in the Kainji Lake National Park in Kwara State, Nigeria. The game reserve is located between 9°50′N and 10°12′N; 4°00′E and 4°34′E (Figure 1). There is a distinct rainy season from April to October with maximum rains in August and September. Generally the mean daily

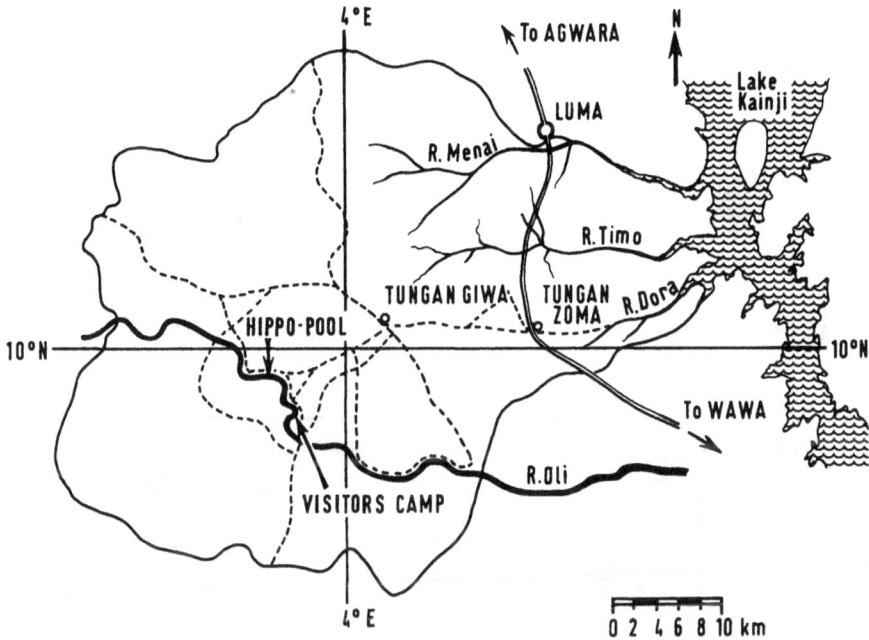

Fig. 1. Kainji Lake National Park, Kwara State, Nigeria.

temperatures are high. The lowest daily temperatures are recorded in December and January during the Harmattan while the highest are recorded in April and May before the onset of the rainy season (Figure 2).

The geology of the area consists of granites, gneisses, migmatites of the Basement complex, schists and some quartzites.

The vegetation of the game reserve is mainly of the Northern Guinea type with some areas showing the characteristics of a transition from Southern to Northern Guinea type. The main vegetation types are *Burkea africana/Terminalia avicennioides* Savannah, *Isoberlinia tomentosa* Woodland, *Terminalia macroptera* Tree Savannah, and *Gardenia/Maytenus* Scrubland (Figure 3).

This experiment was set in the *Terminalia macroptera* Tree Savannah on the River Oli in the game reserve. This vegetation is found in the valleys along the River Oli in seasonally flooded areas with waterlogged soils. This vegetation forms a typical habitat for the kob (*Adenota kob*) where it establishes definite territories. Kob is one of the most abundant large mammals in Borgu Game Reserve with a population of over three thousand. It is also very important for game viewing and tourism. Because of the territorial habits of these animals and also due to their sedentary nature in areas adjacent to water points, it is thought that their grazing impact on the habitat is likely to be significant.

The effects of annual burning on this type of vegetation are not clearly

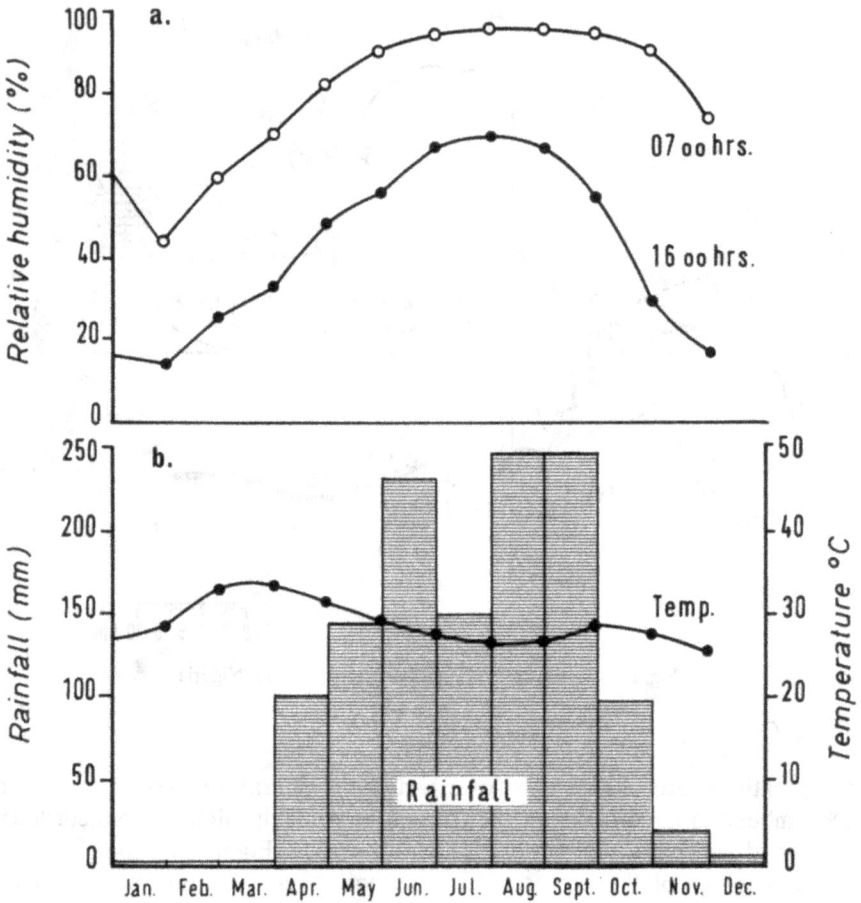

Fig. 2. Monthly averages of rainfall, temperature and relative humidity in the Kainji Lake National Park.

understood. This experiment was then set up to investigate thoroughly the combined effects of fire and grazing on the kob habitats and recommend to the management of the reserve on the best time, season and mode of burning necessary to provide an optimum forage for kob. This experiment is a long term one which will provide the necessary solutions to the burning problems in this habitat in the reserve. In this type of habitat, burning seems to be inevitable especially to maintain good visibility for game viewing, to get rid of last season's litter and to promote a new grass flush for the grazing requirements of the animals. At the same time this burning should be controlled and planned judiciously so as to maintain the necesary cover for the animals, protect the water-sheds, fringing forests and important woodlands and also prevent the destruction of animals by hot fires.

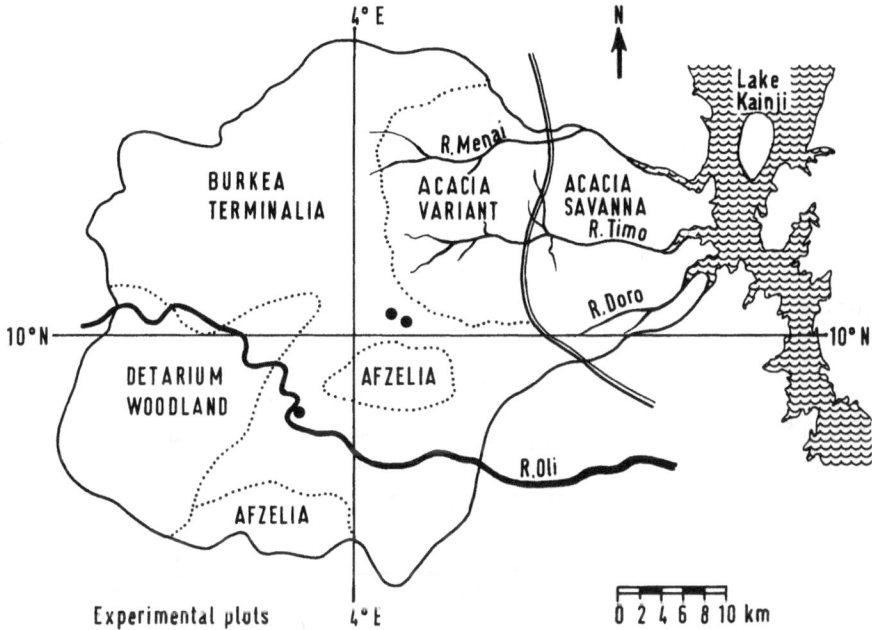

Fig. 3. Main vegetation types in Kainji Lake National Park.

METHODS

The details of the various methods used to collect data from this experiment are fully described by Afolayan (1975).

Data are collected from eight plots (50 m×20 m each) with eight different types of treatments. Each plot is in turn subdivided into four subplots (10 m × 20 m each). The eight different treatments imposed on the habitat are as follows:

(1) Early burning every year with grazing.
(2) Early burning every year without grazing.
(3) Late burning every year with grazing.
(4) Late burning every year without grazing.
(5) Slashing every year with grazing.
(6) Slashing every year without grazing.
(7) Fire excluded with grazing.
(8) Fire and grazing excluded.

The plots from which grazing is excluded are fenced with chain linked wire about 2 m high. There are occasional breakages by elephants and some other big mammals but these are repaired immediately and there is grazing only by rodents and insects.

Measurements of herbage standing crop and litter deposit are made monthly in all the experimental plots. Forty quadrats (1 m² each) are laid at random in each plot and the herbage is clipped in each quadrant. The clipped materials are separated into grasses and forbs and the grasses are sorted into individual species. The litter in each quadrant is also collected. All the various components are weighed. Sub-samples are taken from the various components and oven dried to constant weight, ground and stored for chemical analysis. Soil samples are collected from all plots at random points. Samples are collected at depths of 0–10 cm, and 10–20 cm and these are taken for chemical analysis. The data discussed here are those collected monthly from all plots from August 1974 to January 1975.

RESULTS

Monthly changes in the herbage standing crop and the litter deposits in all the plots are shown in Tables 1 and 2. Confidence limits at the 95 per cent probability level were computed on all results. Figure 4 shows the monthly changes in grass production in relation to rainfall from August 1974 to January 1975.

Maximal grass production was recorded in the months of October and November. A similar situation was noted by Egunjobi (1973) in the moister type of savannah in the Western State of Nigeria. The highest standing crop (1666.49±299.19 gm/m²) was recorded in November in the plot burnt late every year without grazing, while the lowest standing crop (189.29±35.23 gm/m²) was recorded in August in the plot burnt early with grazing. Egunjobi (1973) recorded a maximum standing crop of 1934±169 gm/m² in a plot burnt in January in the derived savannah, where he carried out his experiment. This highest standing crop was attained in September.

Studies on the litter deposit (Table 2) have shown the highest accumulation of litter to be in the slashed and unburnt plots.

DISCUSSION

The preliminary results presented in this paper are not meant for making any snap judgement or rushed conclusions. This is just the first step of a long-term experiment. A more comprehensive account will be written after all the data analysis has been concluded, when sound conclusions and recommendations can be made.

The type of result from any burning treatments on the habitat depends on how the burning is applied. Fire could be a good servant as well as a bad master. Foresters use early burning to eliminate grasslands while range managers will prefer hot fires which would open up bush thickets and produce enough perennial grass for grazing animals. A wildlife manager is also faced with the problem of maintaining cover and protecting water-sheds especially in the dry season and at the same time he tries to open up certain game tracks for better

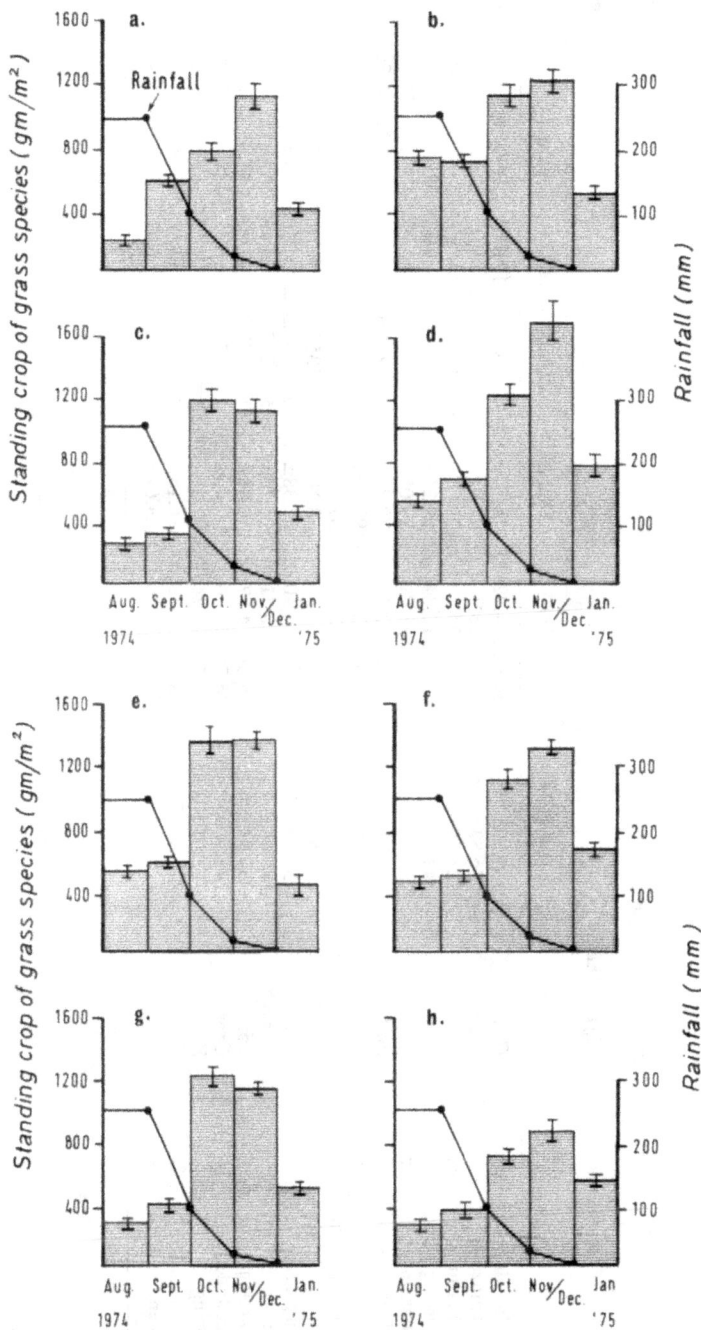

Fig. 4. The monthly changes in grass production in aeration to rainfall from August 1974 to January 1975. See Tables 1 and 2.

TABLE 1

Values of standing crop fresh weights (g/m²) of herbage with 95 per cent confidence limits

Treatments	Type of herbs	Sampling times				
		Aug. 1974	Sep. 1974	Oct. 1974	Nov/Dec. 1974	Jan. 1975
a Early burning every year with grazing	Grass	189.29±35.23	554.88±24.22	778.77± 92.00	1083.57±171.30	398.46± 67.96
	Forb	26.83± 3.05	60.73±13.76	90.02± 31.23	56.35± 11.56	60.24± 17.98
b Early burning every year without grazing	Grass	721.75±61.86	717.38±54.30	1128.85±121.31	1203.67±144.62	513.30± 49.02
	Forb	43.61± 4.19	60.31±10.84	59.26± 10.66	34.58± 9.65	20.72± 3.68
c Late burning every year with grazing	Grass	297.00±17.22	358.38±28.01	1159.05±128.75	1082.14±130.37	450.78± 72
	Forb	81.71±15.22	106.78±12.40	179.24± 32.17	228.44± 38.84	152.62± 48.29
d Late burning every year without grazing	Grass	549.25±41.56	708.63±12.51	1192.99± 72.41	1666.49±199.19	768.10±131.52
	Forb	112.50±12.76	119.79±22.63	137.12± 8.27	66.67± 6.80	137.00± 23.44
e Slashing every year with grazing	Grass	499.58±39.57	588.38±32.38	1338.10±163.85	1351.85±100.52	414.72±119.32
	Forb	34.86± 8.36	38.53±26.40	87.38± 26.40	58.33± 20.21	38.75± 7.96
f Slashing every year without grazing	Grass	435.68±51.57	481.25±40.71	1095.72±116.06	1289.55± 68.48	667.95± 76.61
	Forb	80.88±14.25	115.36±14.67	116.88± 13.61	73.33± 10.00	97.83± 18.25
g Fire excluded with grazing	Grass	272.50±29.40	376.75±38.06	1203.46±123.02	1124.56± 67.25	481.88± 53.02
	Forb	62.65± 3.03	147.59±23.74	237.29± 92.61	182.46± 48.73	81.67± 20.36
h Both fire and grazing excluded	Grass	220.85±20.23	348.13±56.53	686.18± 80.24	845.41±139.42	521.63± 88.56
	Forb	120.63±23.44	268.38±34.94	383.64± 83.73	52.10± 4.92	59.38± 9.95

TABLE 2

Values for litter fresh weights (g/m^2) with 95 per cent confidence limits

Treatment	Sampling times				
	Aug. 1974	Sep. 1974	Oct. 1974	Nov./Dec. 1974	Jan. 1975
a Early burning every year with grazing	56.69± 8.40	33.20± 2.24	10.38± 0.85	34.50± 3.19	47.87±11.56
b Early burning every year without grazing	76.88±30.14	39.13±11.54	32.04±11.56	9.94± 2.45	48.49± 9.16
c Late burning every year with grazing	10.24± 1.87	8.24± 0.59	15.31± 5.65	23.31±79	27.63± 6.43
d Late burning every year without grazing	24.25± 7.40	20.25± 3.49	27.25± 2.67	36.50± 8.33	40.00± 4.10
e Slashing every year with grazing	296.75±75.19	212.23±11.19	201.01±38.44	128.74±22.61	151.32±37.81
f Slashing every year without grazing	521.50±90.38	525.25±63.86	226.48±21.22	210.67±39.61	154.19±18.28
g Fire excluded with grazing	269.50±77.00	269.30±29.72	216.08±36.96	184.65±18.90	162.50±16.35
h Both fire and grazing excluded	341.30±33.13	711.57±43.86	366.08±40.31	184.87±14.66	210.48±18.06

173

visibility in game viewing. In areas where ticks and tsetse flies are common, the ranchers would like to use fire to get rid of them while at the same time farmers in their shifting cultivation process need fire for clearing. A range manager or wildlife manager also needs fire for the removal of old litter and debris to allow new grass production, while some desirable grass species can only be maintained by burning. The only solution to the problems of fire are, controlled and planned burning programmes. A long-term burning plan should be backed up with an experiment to determine the optimum conditions of burning in the area concerned. West (1965) has said that the effects of fire on a drier savannah are more severe than its effects on the moister savannah. This shows that one type of approach does not necessarily provide solutions for problems in different areas. Each habitat needs a specific approach for planning and timing its burning and this should be backed up by scientific experiment. In Kainji Lake National Park where this experiment was sited, the maximum grass standing crop was not attained till the month of November, while this was attained in September at Fasola Ranch, in a moister derived savannah where Egunjobi (1973) carried out his own experiment. The highest standing crop attained at Borgu Game Reserve was (1666.49±299.19 gm/m^2) while the highest attained at Fasola ranch was 1934+169 gm/m^2. From these results, it is obvious that marked differences occur between the effects of burning on the vegetation of Kainji Lake National Park in the Northern Guinea Savannah and the vegetation of Fasola Ranch in the Derived Savannah.

While planning for controlled burning programmes the differences in climate, soils and vegetation of different habitats should be recognized and the prevailing conditions in each habitat should determine the type of burning plan to be formulated for it.

ACKNOWLEDGEMENTS

I am highly indebted to Dr S. S. Ajayi and Dr J. B. Hall of the Department of Forest Resources Management for giving me the necessary guidelines for producing this manuscript. I am equally grateful to Professor L. Roche, Dr S. K. Adeyoju and Dr D. U. U. Okali for their unflinching support in providing the necessary facilities for all my field trips.

I also thank the Director and my former colleagues at Kainji Lake Research Project, and the officer in charge of the Game Preservation Unit of Kwara State for giving full co-operation and assistance during my field studies.

REFERENCES

EGUNJOBI, J. K. (1971) Savannah burning, soil fertility and herbage productivity in the derived savannah zone of Nigeria. *Proc. Symp. Wildlife Conservation in West Africa*. (I.U.C.N. Publication.) No. 22, pp. 52–8.

EGUNJOBI, J. K. (1973) Studies on the primary productivity of a regularly burnt tropical savannah. *Ann. Univ. Abidjan, Series E*, No. 6, pp. 157–69.

HOPKINS, B. (1962) Vegetation of Olokemeji Forest Reserve, Nigeria I. General features and the research sites. *Journal of Ecology*, Vol. 50, pp. 559–98.

HOPKINS, B. (1963) The role of fire in promoting the sprouting of some savannah species. *Journal of the West African Science Association*, Vol. 7, pp. 154–62.

HOPKINS, B. (1964) Some observations on savannah burning in the Olokemeji Forest Reserve. *Proc. 1st Nigerian Forestry Conference*, pp. 59–60.

HOPKINS, B. (1965) Observations on savannah burning in the Olokemeji Forest Reserve, Nigeria. *Journal of Applied Ecology*, Vol. 2, pp. 367–81.

MOORE, A. W. (1960) The influence of annual burning on a soil in the derived savannah zone of Nigeria. *7th Trans. International Cong. Soil Sci.*, Vol. 4, pp. 257–64.

ONOCHIE, C. F. A. (1964) An experiment in controlled burning in the Sudan Zone (investigation 224). *Proc. 1st Nigerian Forestry Conference*, pp. 131–55.

RAINS, A. B. (1963) Grassland research in Northern Nigeria (1952–1962). *Misc. Pap. Inst. agric. Res., Samaru, 1.*

RAMSAY, J. M. and INNES, R. R. (1963) Some quantitative observations on the effects of fire on the Guinea Savannah vegetation of Northern Ghana over a period of 11 years. *Soils Afr.*, Vol. 8, pp. 41–85.

RODGERS, W. A. (1970) The effects of fire on Miombo woodland. (Unpublished Ms. Miombo Research Station, Selous, Tanzania).

WEST, O. (1965) *Fire in vegetation and its use in pasture management with special reference to tropical and subtropical Africa.* Mimeogrd. Publ. Commonw. Bur. Past. Fld. Crops.

21

THE EFFECTS OF BURNING AND GRAZING ON THE PRODUCTIVITY AND NUMBERS OF PLANTS IN RWENZORI NATIONAL PARK, UGANDA

E. L. Edroma

Uganda Institute of Ecology,
Lake Katwe, Uganda

INTRODUCTION

For Uganda Langdale-Brown *et al.* (1964) wrote 'There is little non-forested land below 6000 ft which is not burnt over at least once every three years'. Osmaston (1965) stated that the present appearance of Rwenzori National Park is mainly due to fire. The Uganda National Parks have always been a fire risk. An attempt was made to practise controlled burning in all the three National Parks. Rotational burning is enforced in Kabalega Falls National Park and the effects of fire on the vegetation there were published by Spence and Angus (1971). The implementation of controlled burning scheme in the Kidepo Valley National Park has been described by Ross and Harrington (1969a, 1969b) and Harrington and Ross (1974). In Rwenzori National Park controlled burning scheme has not been implemented although there was a 'no-burn' policy in 1965–66 which later proved impracticable. The fire-breaks constructed became overgrown and ineffective. The prevention of fire resulted in accumulation of litter and ultimately in several fierce uncontrolled fires one or two years later. Since the late 1960s fires starting in the Park have been allowed to burn.

In Uganda burning was at first condemned through the influences of the forestry policy. In 1903 rules under the Forestry Ordinance required 'no person shall set fire to any forest, grass or undergrowth on any Crown lands' (LUP 1910). In 1920 the Careless Use of Fire Prevention Ordinance aimed at preventing damage to crops, bush and grass by non-owners of the land, and to permit owners to light fires on their own land provided that they prevented its spreading to other persons' land or to unoccupied land (LUP, 1924).

The campaign for a change of burning from the traditional, fierce, late burn to a less damaging early burn engineered by the Ugandan Forestry Department with the support of the Agriculture and Veterinary Department was at first a failure (A.R.F.D., 1926) and later a success (A.R.F.D., 1937). The result of the policy seems to be a partial increase in the density of bush cover of species with no economic value, a decrease in the value of grazing lands and spread of tsetse flies *Glossina morsitanus* and *G. pallidipes* Aust. (Burton, 1948; Robertson and Bernacca, 1958). From 1950 the foresters showed less interest in influencing the burning policy outside the designated forest reserves. They use early fires to control the spread of their forests, burning woodland edges so as to prevent accidental late fires into woodlands and burning savannah forest reserves in order to encourage thickening of the woody cover (A.R.F.D., 1956), although the operations have been less effective silviculturally. For the reasons mentioned above and because early fires can expose the bare ground to strong environmental forces and often offer no, or minimal, herbage for the wild animals for a long time during the dry season—a decree was signed in 1974 which prohibits grassland burning early in the dry season and allows fires only under the direction and supervision of a scientific officer. The work reported here attempts to examine the influences of burning on the dry matter production and changes on the numbers of plants in grazed and protected *Hyparrhenia–Themeda* grassland.

THE STUDY AREA

Rwenzori National Park is situated in western Uganda between 29°45′E, 30°15′ E and 30′S and 0° 15′ N. It lies in the floor of the western arm of the rift valley and is surrounded by the Rwenzori horst to the north and Lakes Edward and George to the west and north-east respectively, while to the north-west lie the escarpments of the western rift. The park receives, on average, 884 mm of rainfall a year, predominantly during two wet seasons (March–May and August–November). December–February and June–early August are dry periods. The vegetation of the Park has been described by Edroma (1975) and the study area to which this paper refers is rich in *Hyparrhenia–Themeda* 'fire-climax' grassland and has fertile, black, loamy soil. This type of grassland is the main grazing ground and supports a large variety of wildlife ungulates.

The grazed and enclosed study area was divided into three parts and early, late, and no burn treatments were applied to them annually from 1971. Early burning was carried out at the beginning of the dry season, usually in the third or fourth week of June. The late burn was set at the end of the dry season just before the rains started — usually towards the end of August. Selective removal of certain grass species in permanent 1 m² quadrats was superimposed in order to simulate the removal of some plants by elephants and other ungulates. These were (1) no vegetation removal (control) (2) removing the dominant species (*Hyparrhenia filipendula* Stapt) leaving the other plants, and (3) removing all other species leaving the dominant one. The removal every February, June, August and December involved uprooting of the entire plants with the minimum amount of disturbance caused to the soil surface or to the remaining vegetation. At each observation, the numbers of plants per species removed or left intact were recorded.

Three 1 m² sampling plots were established in each plot in early June 1973. One year later, the standing crop in the quadrats was harvested at ground level and sorted out into the main two species and the remaining plants were grouped together. The herbage was dried and weighed to assess the net primary production in the fire treatments.

RESULTS

Dry matter production

The growth of *H. filipendula* and *Themeda triandra* Forsk was accelerated in the late burn plots. *H. filipendula* produced 51.2 per cent and 40.1 per cent while *T. triandra* 29.6 per cent and 48.1 per cent of the herbage in the grazed and fenced plots respectively. In the grazed early burn plot, *H. filipendula* produced more (57.4 per cent) and *T. triandra* less (21.5 per cent). But the order was reversed in the enclosed plot when *H. filipendula* produced less (38.1 per cent) and *T. triandra* more (51.9 per cent). When fire was excluded growth of these species was invariably reduced. The number of species other than them was most in the unburnt and least in the burnt plots. So burning appeared to favour these two species. Late burn plus grazing increased the production of *T. triandra*, but decreased that of *H. filipendula*; without grazing, late burn was more favourable to *T. triandra* than to *H. filipendula*. Late burn therefore appeared to encourage greater production of *T. triandra* than of *H. filipendula*, while early burn together with grazing generally promoted more *H. filipendula*.

Development of numbers of plants

No burn

When the dominant *H. filipendula* was removed, the numbers of *T. triandra*, *Bothriochloa insculpta* A. Rich, *Microchloa kunthii* Deav., *Sporobolus stapfianus* Gand, *Sporobolus pyramidalis* Beauv., *Aristida adoensis* A. Rich, *Chloris pycnothrix* Trin and *Digitaria scalarum* Chiov. all increased. The rate of increase in each species was higher in the enclosed, and lower in the grazed, plots. With the exception of *C. pycnothrix*, these species multiplied in numbers during the first fifteen months of removing the dominant *H. filipendula* from the sward. In the second half of the period, their numbers became relatively stable. More seedlings of *H. filipendula* were removed every three months in the enclosed than grazed plots. Where all the species were removed leaving *H. filipendula* alone, the number of *H. filipendula* was boosted higher in the protected than in the grazed plots. The number of *T. triandra*, *S. pyramidalis*, *S. stapfianus*, *B. insculpta*, *M. kunthii*, *D. scalarum* and forbs all increased in the plots particularly in the grazed area. In the control plots, the numbers of *H.*

filipendula, B. insculpta, Heteropogon contortus Roem and Schult and *Chloris gayana* Kunth slightly built up inside but decreased outside the enclosure. The numbers of *T. triandra* dropped by 40.3 per cent without, and by 60.1 per cent with, grazing during 1971 and 1973. These reductions in the grazed plot were attributed to grazing and those in the enclosed plot to fire exclusion. On the other hand, the numbers of *S. pyramidalis* and *S. stapfianus* lowered in the control plots where the vegetation cover was too thick for them. The augmented occurrence of *C. pycnothrix, A. adoensis, D. velutina, S. pyramidalis, S. stapfianus, M. kunthii, Eragrostis cilianensis* (All.) and *Euphorbia* species in the plots receiving removal of the dominant species or, to a less extent the removal of all the species leaving the dominant one, suggests that these treatments created the habitat favourable to the species just mentioned, which are otherwise usually found in degraded habitats. Secondly the greater increase of their numbers in the quadrats where the dominant *H. filipendula* was removed rather than in the control, suggest that the dominant species was suppressing them and reducing their niches in the natural sward. The higher numbers of plants in the grazed plots where all the species were removed leaving the dominant *H. filipendula* points out the influence of the animals in creating conditions favourable for seed germination and seedling establishment.

Early burn

When the dominant *H. filipendula* was removed, the numbers of *B. insculpta, S. pyramidalis, M. kunthii, Tephrosia* and forb species all increased in the plots, but significantly in the grazed than ungrazed plots. Those of *T. triandra, A. adoensis* and *Indigofera circinella* Bak. rose in the enclosed plot, but remained unchanged in the grazed area throughout the observation period. Fewer seedlings of *H. filipendula* were recorded at each vegetation removal, decreasing with time at a higher rate in the grazed than in the fenced plots. Removal of all these species leaving the dominant *(H. filipendula)* resulted into a change in the number of *H. filipendula* by a maximum increase of 20.2 per cent in the non-grazed plot and a maximum drop of 25.5 per cent in the grazed area between 1971 and 1973. Moderate numbers of seedlings of *T. triandra* and *B. insculpta* and considerable amounts of seedlings of *S. pyramidalis, M. kunthii, A. adoensis* and forbs were recorded in the quadrats. Appreciable numbers of *E. cilianensis* appeared in the grazed plots. In the control plots, more plants of *H. filipendula, B. insculpta, M. kunthii, S. pyramidalis* and *H. contortus* were recorded in the protected compared to those from the grazed plot. The number of *H. filipendula* decreased significantly from the second burning in August 1972 onwards. *T. triandra* maintained stable populations in both the grazed and protected plots, but with a rising trend inside the enclosure. *B. insculpta* steadily decreased in number from an average of 11.0 to 0.3 in the grazed plot between June 1971 and August 1973. The numbers of *S. stapfianus* fluctuated considerably being high in the wet seasons and low in the dry seasons. The numbers of forbs were largest in the treatment where all the species were removed leaving *H. filipendula*, the lowest in the control. More species were recorded in the quadrats where the dominant species was removed and less in the plot where all the species were removed leaving *H. filipendula*.

Late burn

Removal of *H. filipendula* resulted in the remarkable appearance of seedlings of *H. filipendula* and of plants of *T. triandra, B. insculpta, S. pyramidalis, M. kunthii* and *I. circinella*. The numbers of *C. gayana, H. contortus* and *A. adoensis* increased in the grazed plot but were absent (*H. contortus* and *C. gayana*) or few *(A. adoensis)* inside the enclosure. *Brachiaria decumbens* Stapf. increased inside the enclosure, but it appeared outside the enclosure nearly two years of the fire treatment. The other species which appeared in the plots with several numbers of individuals were *Cassia mimosoides* from *G. sensu* Brenan., *Polygala erioptera* DC., *Setaria sphacelata* Stapf. and L. E. Hubb *C. pycnothrix, Brachiaria platynota* (K. schum.) Robyns and species of *Alysicarpus*. Species of *Tephrosia* and forbs generally decreased in this fire treatment. When all the species other than the dominant species were removed, the number of *H. filipendula* went up by 19.9 per cent in the ungrazed plot, but slightly declined by 0.7 per cent in the grazed plot between June 1971 and June 1973. Of the plants repeatedly removed those of *T. triandra, S. pyramidalis, B. insculpta, M. kunthii* and forbs continued to re-establish in considerable numbers and sixteen other rare species persisted in the plots. *H. filipendula, T. triandra* and *M. kunthii* maintained stable populations. In the control, more *H. filipendula* and *T. triandra* were recorded in the protected than in the grazed plots, whereas more *M. kunthii* was observed in the grazed than ungrazed areas. The number of *S. pyramidalis* increased in the grazed plot, but remained virtually stable in the ungrazed plot. *S. stapfianus* dropped by 43.4 per cent in the enclosed and by 23.3 per cent in the grazed plots. Generally, fewer plants and species were observed in the grazed than protected plots.

DISCUSSION

Herbage production

Burning had a marked effect on the potential for dry matter production. Whether grazed or not, burning stimulated growth of the grassland, and the species other than the dominant ones were promoted to contribute more in the total herbage production. The standing crop was higher in the burnt than unburnt plots despite the fact that the fire removed the old growth. The early and late burning regimes had very different effects on the productivity of the grassland. The maximum standing crop was recorded in the late burn and the minimum in the no burn plots. The fire removed coarse and fibrous dead shoots which were of no direct food value to the animals. *T. triandra, H. filipendula* and the other desirable species were maintained productive. Within one or two weeks after the late burn, a flush of leguminous seedlings resulted. In the early burnt plots few leguminous seedlings emerged nearly two months later during the rainy season. The stimulation of dry matter production of fire is not unique to Rwenzori National Park (e.g. reviews by West (1965) for Africa, Ahlgren and Ahlgren (1960) and Daubenmire (1968) for America). Cushwa *et al.* (1968) reported six times more grasses and two times more leguminous plants on burnt than unburnt controls. Kucera and Ehrenreich (1961) noted significant increases in grassland productivity on burnt Central Missouri Prairie. Burning has been found to reduce productivity in western North Dakota (Dix, 1960); but Daubenmire (1968) suggested that such reductions can be correlated with years of deficient precipitation and aggravated by the tendency of the soils to dry out more quickly on the burnt areas. A similar conclusion is reached in some parts of southern Africa, that burning increases grassland production in moist regions, but is detrimental in more arid areas (Staples, 1945; West, 1965). But it seems to be problematic, if this conclusion is generally applied, since fire influences are found to be detrimental in many vegetation studies.

Two dry seasons per year in Rwenzori National Park are fairly short (some 2 months each) to cause total drying up of the grasses and destructive conflagration of late fires. The late fires in the study areas are not as violent as those found in north Uganda (Harrington and Ross, 1974) and many parts of West Africa (Knapp, 1973) where the dry seasons are long (at least 4 months) and the late fires may be highly destructive to the vegetation. The short exposure of the soil and the regrowth to grazing throughout the dry period in Rwenzori National Park causes adverse effects on the recovery of the vegetation. For species like *T. triandra* which are sensitive to heavy grazing (Edwards and Bogdan, 1951) a grazed early burn would be damaging. Referring to African grasses, Botha (1945) and Cook (1939) advocated regular burning in order to maintain best quality of forage of desirable *T. triandra*. Also in the United States, burning seems to improve quality and quantity of herbage in certain prairie communities (Hensel, 1923; Campbell *et al.*, 1954; Duvall, 1962; Grelen and Epps, 1967).

Influences on numbers of plants

Vegetation removal of one or more species was followed by the establishment of seedlings of the species removed and of other species which had not been in the plots previously. *C. pycnothrix, E. ciliaris, Panicum brevifolium* L, etc. were some of the species which were absent at the start of the experiments and they invaded the plots from which some component species were removed regardless of the burning regime. These species are otherwise commonly found in the overgrazed grasslands where the vegetation cover is poor. They are normally absent in the tall grasslands. The vegetation removal reduced the normal foliage cover of the grassland and created the habitats suitable for their establishment. The kind of the establishment suggests that the soils of the park contains seeds of a variety of species in the region, being remnant of former vegetation or brought in probably by wind and animal dispersion, and that they remain viable in the soil presumably for a long time (see Roberts, 1963; Numata *et al.*, 1964).

Removal of grasses and dicotyledonous species were shown to increase the seedling establishment of certain species (Putwain and Harper, 1968; Knapp, 1954; Sagar, 1959; Harper *et al.*, 1965). The large increase in numbers of seedlings of the species removed and of new species, suggest that the species have reduced overlapping niches which expand when certain components of the community are removed. They also indicate the powerful role of the grasses in limiting the realized niches of the forbs. In the present study, vegetation removal increased the open area, reduced competition, allowed more light to reach the ground surface and increased microsites for seed germination and seedling establishment. Concerning wildlife influences, uprooting of plants by grazers like hippos and elephants, and scooping of the soil surface by animal hoofs create microsites. Defoliation of the vegetation by grazing also checked the thickness of the canopy and thereby allowed more solar energy to reach the soil surface. Increase of soil temperature by such rise in light energy and/or by fire are some of the factors which in many cases trigger seed germination.

Selective removal of the dominant species resulted into an increase in the amounts of a large number of species. The increases were particularly noticeable in the late burn plots and relatively less in the unburnt and early burn plots. In the unburnt controls receiving no vegetation removal, *H. filipendula* increased inside but decreased outside the enclosure. *T. triandra* decreased by 60 per cent in the grazed and by 40 per cent in the ungrazed plots in three years. The number of the dominant species increased in the late burn treatment. When one plant died, its exact place remained unoccupied throughout the study period, but seedlings of the same species or other species established near the stumps of the dead plants. In this way, the amounts of individual plants maintained stability in the swards. A large proportion of the seedlings died before reaching maturity. These changes in the number of seedlings in the grassland reflect the findings of Harper and McNaughton (1962) and of Knapp (1954) that mortality is highest in the seedlings stage and of Tamm (1948) and Kershaw (1962) that older plants are better competitors.

Burning favoured practically all the main species in the grassland with the exception of *Panicum maximum* Jacq. Late burning had a decisive effect on the botanical composition of the grassland. The main effect of burning was to give dominance of *T. triandra* and this presumably arose from the established plants being resistant to fire damage and from changes in the microclimate near the soil which favoured germination of their seeds and reduced the ground cover of other species. The high number of legume seedlings following fire may be explained by the suggestion of Cushwa *et al.* (1968) that moist rather than dry heat greatly increased both germination rate and total germination of legume seeds. It may well be that fire and the resulting ash prepared a favourable seed bed (see Knapp, 1974) allowing a better chance to germination in nutrient-rich soil. The new habitat presumably reduces seed and seedling destruction by other organisms. Another possibility is that water as one of the main products of burning, condenses on cooler objects like seeds. Legume seeds are small with hard testas (Kawatake *et al.*, 1955; Amen, 1963) and the increased temperature and moisture following fire may break dormancy and stimulate them to germinate in abundant profusion.

The tussocky species (e.g. *Sporobolus*) performed better under early than late burns. Being short plants up to 10 cm in height, the hotter fires of the late burns were probably too severe and damaging to their shoots and roots. These plants are commonly found in the *Sporobolus* and mosaic grasslands where fires are nearly excluded by the absence of vegetation cover as fuel. They therefore seem not to be adapted to the same degree to fires as the tall grasses (e.g. *T. triandra, H. filipendula*). Intolerance of shrubs and trees to late fires and their preference for early fires are no exception in the grassland of Rwenzori National Park.

West (1965) stressed the importance of litter and concluded that burning favoured *Themeda* because it removed litter rather than as a direct fire effect. The untrampled litter in the unburnt enclosed plot accumulated several centimetres above ground. The encouragement of *T. triandra* by burning is of significance in the park management policy. The present study provides data where seed germination and plant vigour were better in the late than early, and least in the no burn treatments. Burning seems to be cheapest and best practice for promotion of *T. triandra* and its associate species.

SUMMARY

In Rwenzori National Park, Uganda, burning stimulated herbage production more in the late than early burn regions. The maximum standing crop was recorded in the late burn and the minimum in the no burn plots. It increased species diversity and quality of herbage and therefore utilization of the range by the wild animals. Early burning followed by heavy grazing was found to be detrimental to most species and it caused habitat deterioration. Different species showed differences in response to different fire regimes and grazing intensities.

Selective removal of one or more species in a community resulted into establishment of seedlings of the species removed and of other species which had not been in the sward previously, and into cohabitation of the area by a large number of species. The increases were more in the late burn and less in the unburnt and early burn plots. When a plant died its exact place remained unoccupied, but seedlings of the same species or other species established near the stumps of the dead plant. Stability in the swards was thus maintained.

REFERENCES

AHLGREN, I. F. and AHLGREN, C. E. (1960) Ecological effects of forest fires. *Botanical Review*, Vol. 26, pp. 483–533.

AMEN, R. C. (1963) The concept of seed dormancy. *American Scientist*, Vol. 51, pp. 408–28.

A.R.F.D. (1926) Annual Report of the Forest Department, Entebbe.

A.R.F.D. (1937) Annual Report of the Forest Department, Entebbe.

A.R.F.D. (1956) Annual Report of the Forest Department, Entebbe.

BOTHA, J. P. (1945) Veld management in the eastern Transvaal. *Farming in South Africa*, Vol. 20, pp. 537–41.

BUXTON, P. A. (1948) *Trypanosomiasis in Eastern Africa*, H.M.S.O. London.

CAMPBELL, R. S., EPPS, E. A., MORELAND, C. C., FARR, J. L., and BONNER, F. (1954) Nutritive values of native plants on forest range in Central Lousiana. *La. Agric. Expt. Stat. Bull.*, Vol. 488, p. 18.

COOK, R. (1939) A contribution to our information on grass burning. *South African Journal of Science*, Vol. 36, pp. 270–82.

CUSHWA, C. T., MARTIN, R. E. and MILLER, R. L. (1968) The effects of fire on seed germination. *Journal of Range Management*, Vol. 15, pp. 250–4.

DAUBENMIRE, R. F. (1968) Ecology of fire in grasslands. *Advances in Ecological Research*, Vol. 5, pp. 209–65.

DIX, R. L. (1960) The effects of burning on the mulch structure and species composition of grasslands in Western North Dakota. *Ecology*, Vol. 41, pp. 49–56.

DUVALL, V. K. (1962) Burning and grazing increase herbage on slender bluestem range. *Journal of Range Management*, Vol. 15, pp. 14–6.

EDROMA, E. L. (1975) The influences of burning and grazing on productivity and population dynamics of grasslands in Rwenzori National Park, Uganda. (Ph.D. thesis, University of Giessen.)

EDWARDS, D. C. and BOGDAN, A. V. (1951) *Important grassland plants of Kenya*, Nairobi.

GRELEN, H. E. and EPPS, E. A. (1967) Herbage response to fire and litter removal on southern Bluestem range. *Journal of Range Management*, Vol. 20, pp. 403–4.

HARPER, J. L. (1967) The regulation of numbers and mass in plant population. In the Proceedings International Symposium on Population Biology and Evolution.

HARPER, J. L., WILLIAMS, J. T. and SAGAR, G. R. (1965) The behaviour of seeds in soil. I. The heterogeneity of soil surfaces and its role in determining the establishment of plants from seed. *Journal of Ecology*, Vol. 53, pp. 273–86.

HARRINGTON, G. N. and ROSS, I. C. (1974) The savanna ecology of Kidepo Valley National Park, Uganda. I. The effects of burning and browsing on the vegetation. *East African Wildlife Journal*.

HENSEL, R. L. (1923) Recent studies on the burning on grassland vegetation. *Ecology*, Vol. 4, pp. 183–8.

KAWATAKE, M., TAKIZAWA, M. and KATAYAMA, M. (1955) Studies on the developmental processes of hard seed of some legumes. *Div. Plant Breed. and Cult. Bull.*, Vol. 30, pp. 177–182; pp. 193–202.

KERSHAW, K. (1962) Quantitative ecological studies from Landmannhellu, Iceland. *Eriophorum engustifolium. Journal of Ecology*, Vol. 50, pp. 163–9.

KNAPP, R. (1954) *Experimentelle Soziologie der hoheren Pflanzen*. (Verlag Eugen Ulmer, Stuttgart). P. 202.

KNAPP, R. (1973) *The vegetation of Africa with references to Environment, Economy, Agriculture and Forestry Geography*. (G. Fischer, Stuttgart).

KNAPP, R. (1974) *Handbook of vegetation science*. Part VIII. Vegetation dynamics.

KUCERA, C. L. and EHRENREICH, J. H. (1962) Some effects of annual burning on Central Missouri prairie. *Ecology*, Vol. 43, pp. 334–6.

LANGDALE-BROWN, I., OSMASTON, H. A. and WILSON, J. G. (1964) *The vegetation of Uganda and its bearing on land use*. (Govt. Printer, Entebbe.)

L.U.P. (1910) *Laws of Uganda Protectorate*. Ed. Ennis, G. F. M. and Carter, W. M., p. 450.

L.U.P. (1924) *Laws of Uganda Protectorate*. Ed. Griffin, C., p. 1088.

NUMATA, M., HAYASHI, I., KOMURA, T. and OKI, K. (1964) Ecological studies on the buried-seed population in the soil as related to plant succession. I. Japan. *Journal of Ecology*, Vol. 14, pp. 207–15.

OSMASTON, H. A. (1965) The vegetation. In *The Uganda National Parks Handbook*.

PUTWAIN, P. D. and HARPER, J. L. (1968) Studies in the dynamics of plant populations. II. Components and regulation of a natural population of *Rumex acetosella* L. *Journal of Ecology*, Vol. 56.

ROBERTSON, A. G. and BERNACCA, J. (1958) Game elimination as a tsetse control measure in Uganda. *East African Agricultural Journal*, Vol. 23, p. 254.

ROBERTS, H. A. (1963) Studies on the weeds of vegetable crops. II. Effect of six years of cropping on the weed seeds in soil. *Journal of Ecology*, Vol. 50, pp. 803–13.

ROSS, I. C. and HARRINGTON, G. N. (1969a) The practical aspects of implementing a controlled burning scheme in the Kidepo Valley National Park.

ROSS, I. C. and HARRINGTON, G. N. (1969b) The practical aspects of implementing a controlled burning scheme in Kidepo Valley National Park (Second year of operation). *East African Wildlife Journal*, Vol. 6, pp. 101–5.

SAGAR, G. R. (1959) The biology of some sympatric species of grassland. (Unpubl. D. Phil. Thesis, University of Oxford.)

SPENCE, D. H. N. and ANGUS, A. (1971) African grassland management; burning and grazing in Murchison Falls National Park, Uganda. *Symposium of the British Ecological Society*, Vol. 11, pp. 319–31.

STAPLES, R. R. (1945) Bush and different grazing as measures to improve pastures. *East African Agricultural Journal*, Vol. 10, pp. 217–22.

TAMM, O. (1948) Observations on reproduction and survival of some perennial herbs. *Botaniska Notiser*, Vol. 3, pp. 305–21.

WEST, O. (1965) Fire in vegetation and its use in pasture management with special reference to tropical and sub-tropical Africa. (Mimeogrd. Publ. Commonw. Bur. Fld. Past. Crops, 1/1965.)

SECTION 4
MANAGEMENT, TRAINING AND EDUCATION

22

WHAT IS WILDLIFE MANAGEMENT?

C. A. Spinage

College of African Wildlife Management,
Mweka, Moshi, Tanzania

INTRODUCTION

A consideration of wildlife management in many countries suggests that management is seldom taken seriously until animal stocks are depleted to the point of near extinction. Until this state is reached *laissez-faire* policies of 'non interference' are generally adopted in the belief that the animals can look after themselves. Parts of West Africa fall into the category of having sadly depleted stocks of wildlife and we may suppose, therefore, that management in those areas will wish to embrace the most sophisticated approaches.

The concept of wildlife conservation is not new in West and West Central Africa; Phillips (1935) lists at least 39 game reserves in existence in 1935, commencing with the Kwahu Reserve in Ghana dating from 1911. Chad has the oldest national parks, with parks dating from 1933 and 1934. But despite this early association with fauna preservation, either concepts of *laissez-faire* management, or maladministration, have meant that most of the national parks and game reserves in West Africa today are regarded as being in an early stage of development, and the region has fallen behind in preservation and management. This should have the advantage that the region can now benefit from the experience of the eastern and southern parts of Africa without having to undergo their trial and error approaches, and without repeating their mistakes. But first we must define what we are trying to do, and this brings us to the question: 'What is wildlife management?' The answer is that African wildlife management cannot be easily defined, for its application embraces many different situations. The word 'management' can be defined as the manipulation or skilful handling of a resource but we can consider management beginning when we define an area on a map. It is unlikely that such an area will be so large that we do not immediately introduce artificial influences; for such a line drawn on a map, whether around a game reserve or a national park, almost always implies some change in land use, or change in intensity of land use, outside of the area. There may be increased settlement — opening up to hunting, or even the continuation of hunting with its cessation at the boundary line; development of ranching, and so forth.

This limits the ecosystem of the inhabitants within the boundary and is almost certain to effect changes: although nothing at all may be done within the delineated area. But we generally understand management to imply more than the drawing of a line on a map; it usually embraces some form of active manipulation of the biota. I have found it convenient to think of African wildlife management under two major divisions, which I have termed the primary division of management and the secondary division of management. These categories each have two sub-divisions; the former can be divided into management of wildlife within and outside national parks. The second can be divided into management for ecological integrity, and management for exploitation (see Figure 1).

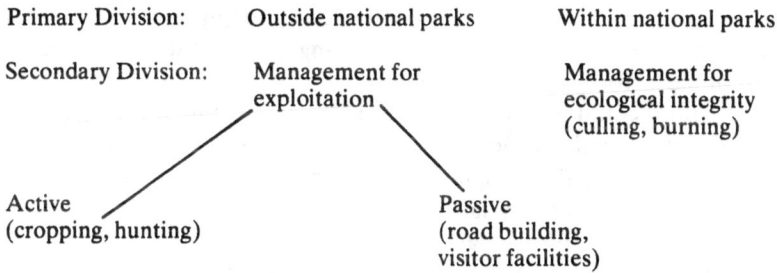

Primary Division: Outside national parks   Within national parks

Secondary Division: Management for   Management for
                    exploitation   ecological integrity
                                           (culling, burning)

Active   Passive
(cropping, hunting)   (road building,
                                 visitor facilities)

Fig. 1. The divisions of African wildlife management.

THE PRIMARY DIVISION OF MANAGEMENT

Management of wildlife outside of national parks

This aspect of management usually relates to game reserves, of which there are various categories, but which all have some legal protection for the animals. We have seen that they have been established in West Africa since at least 1911; and in East Africa they have been in existence since 1900, while the Sabi Reserve in South Africa, now the Kruger National Park, was formed in 1898. Such early reserves and their attendant laws usually implied a restriction on hunting without a licence, which was seldom enforceable; but at a later date some implied a curtailment of human activity in the form of settlement. The concept behind the game reserve was primarily that of preserving species by preventing too many being hunted.

No thought was given to the perpetuation of the species by ensuring optimum conditions for its survival. And, indeed, why should there have been? For there was all Africa with its great herds of game. It was assumed that a species would surely survive if not too many were killed.

The first 'sanctuary' — an area where certain species are afforded total protection from hunting — was established in East Africa around Moshi in 1896. Here, it was forbidden to hunt zebra, eland, antelopes, giraffes, buffalo, ostriches and secretary-birds by the Ordinance of the Imperial Governor, 1896. This did not ensure the survival of any of these species in that area; although the last being very mobile can probably be still found there. Much earlier protective legislation was passed in South Africa, the mountain zebra was protected in an area north of Cape Town in 1780. A few years ago it was estimated that only 75–80 of this species remained.

Management of wildlife in national parks

The first national park in Africa was created in the former Belgian Congo in 1925, and embodied the concept of preserving sectors of natural Africa from man-made change primarily for scientific interest. The park was established by Belgian Royal Decree which commenced with the words 'The Park is created with a scientific objective' (Le Parc est crée dans un but scientifique), but this original aim for African national parks has long been lost sight of; instead, this concept of parks forming 'nature laboratories', areas in which the indigenous flora and fauna could be studied in the absence of human interference, has changed with time in deference to the American idea that such areas should be for the enjoyment of the people, a place for leisure rather than for serious study. Indeed, the tendency is to debate in some circles whether research should be allowed in national parks at all.

Given that an area was set aside from human interference, it was once thought, and still is among some, that it would be ecologically self-perpetuating; that is, continuing in the state in which it was set aside. In other words, no management would be necessary. The first Belgian national parks administrators were not, however, blind to the possible need to manage, and the desirability or otherwise was debated in their first five-year report (Premier Rapport Quinquennial 1935–39). Nevertheless they erred on the side of allowing nature to 'take her course'; a *laissez-faire* attitude relieving the administrative body of responsibility, which was later readily adopted by many other national parks systems. What was not appreciated was that such protected areas would become ecological islands, and that none would be large enough to be self-perpetuating in the state in which they were set aside. Examples of these protected but changing areas are now piling up from many parts of Africa.

THE SECONDARY DIVISION OF MANAGEMENT

Management for ecological integrity

This concerns the management of an area and its animals to prevent changes, induce changes, or to deflect changes, in the ecosystem. This is primarily a preoccupation of national parks. The implications have caused much controversy in recent years concerning the extent of management to be applied, termed interference by the strict conservationists. But the definition of conservation is 'to keep from decay, change, or *destruction*' and this must imply management. To adopt a policy of non-intervention when changes are taking place, by definition cannot be termed conservation. This conflict of thought has led to procrastination and vacillation on almost every important management issue which has confronted East African national parks.

Apart from pressure on the land for grazing and settlement in some areas, probably the two most important management problems confronting national parks in East Africa todday, are those relating to changed wrought by fire and by elephants; problems which have their parallels in West Africa. In the case of fire, areas may have been set aside as parks in a 'deflected climax' state: for example, grassland, carrying high populations of grazing animals, which has been maintained by burning. If burning is prevented then the area may revert to bush and thicket, and the animals disappear. Despite some 70 years of discussion and nearly half-a-century of sporadic research, we still seem to know very little about fire and its relationship to woody growth. This is because the interaction depends upon rainfall, soil and species, which differ from area to area. Thus most, if not all, wildlife managers, do not know whether to burn or not, or even when to burn. In the majority of cases they have continued to burn out of necessity, rather than from understanding, to prevent a build up of combustible material.

But changes due to fire are concealed from the inexperienced eye. One may have a dense, healthy, fire-maintained grass sward, which is, nevertheless, unpalatable to grazing animals for much of the year.

Changes wrought by too many elephants, on the other hand, are extremely obvious and more pronounced and potentially more dangerous to the stability of the ecosystem, the drier the area. The issue which arouses so much feeling is, should elephants in protected areas be destroyed in order to prevent them wreaking habitat change? Antagonists of the policy of reduction consider such habitat changes as may be occurring to be natural and that they should be allowed to run their course. The fact that they are the direct result of man's interference by arbitrarily declaring the area 'protected', is ignored. Management in this case devolves not upon the questions of 'how many?' or 'what age groups?' should be eliminated (and surely know sufficient of population processes to make informed judgements in these respects) but simply that of deciding upon a policy in relation to an area, and pursuing that policy. It is nothing new for feelings to run high on this matter. When the Cape Province administration in South Africa decided to eliminate the troublesome elephants inhabiting an area known as the Addo in 1919 (Bush in 1924) the exercise was called off before completion 'due to the chorus of protest from the whole world', as one writer put it. This was after what was considered the incredible slaughter of some 80 elephants. Nowadays we need speak in thousands when we talk of elephant elimination.

What management in a national park means in this respect is manipulating the animals in order to maintain the integrity of the ecological island. Settlement outside of the boundary, restriction of seasonal movements outside, threatened settlement within the protected area; these are all political problems, which will influence the extent of management within the protected area. But it is within the area itself that the manager must be concerned. If a seasonal refuge is denied to a population, then the wildlife manager must manipulate the animals, perhaps by reducing their numbers, in order to adapt to this change and thus preserve the integrity of his ecological island.

Culling, the planned reduction of a population, does not have as many inherent difficulties in its application as does cropping — taking a sustained yield from a population. We need to be sure of removing enough animals so that we do not initiate a rapid response to increase; but we must avoid killing so many that the population goes into a continued decline. There seem to be no equations formulated for this, but the reduction of post-pubertal to middle-aged females is the quickest way to slow up population increase. Culling is now admitted to by national parks, but what has traditionally been termed 'game control' has for many years been practised outside such areas. It has always, in such circumstances, been badly applied, lacking any objective basis for the selection of, and numbers of, animals killed.

The opposite to culling is the encouragement of low density populations, which might come under 'management for exploitation' if the aim is to increase visitors' enjoyment by presenting an animal spectacle. However it might also fall under the concept of repair of ecological integrity of an area.

It is an aspect of management that we know little about, partly because researchers avoid studying an animal which is thin on the ground: it may not produce enough information for academic attainment or justification of the time spent on it. We have the example of the very successful increase of the white rhinoceros in South Africa — but seemingly no information as to why the animals responded in the way that they did. Some species seem to have difficulty in maintaining their numbers, probably because of environmental factors such as enzootic diseases, predation, food or cover. Others tend towards extinction if a certain minimum population density is reached, perhaps owing to a lack of opportunities for mating. It is an aspect of management which is sadly deficient in knowledge, especially when one considers the interest which is shown in vanishing species.

Management for exploitation

In this category of management we are concerned with two types of exploitation: the passive exploitation of the fauna of an area for man's enjoyment, and the active exploitation of the fauna of an area by sport hunting or cropping. The former can best be described as estate management. In a national park this entails the building of roads and other facilities for human access and use of the area. In hunting areas roads and visitor facilities are less important. The former development needs no elaboration except to say that park planning has recognized approaches which are frequently not adhered to.

It should not, for instance, be the prerogative of a warden in charge of a park to put in roads and tracks where he fancies. They should be aligned according to the best principles of park planning.

The active exploitation of the fauna of an area is usually the work of game departments, the agencies responsible for animals outside of national parks. It is fair to say that the management of animal populations for hunting is virtually non-existent in East Africa, and probably most other parts of Africa also. In legitimate sport hunting we understand it to mean seeking the finest-looking animals. In bovids those with the longest horns, in elephants those with the biggest tusks, and so on. For this purpose game areas may be divided into sectors, which are then booked for hunting. If it is reported that a particular sector is not providing the required trophies, then that sector may be closed for some arbitrary period of time (quite unrelated to any knowledge of recruitment rates), until it appears to be useful again for exploitation. So the only preservation of the resource is that of preventing its entire elimination. Management in this respect should relate to knowing the number of makes which can be shot in an area, based upon a knowledge of population size and structure: and the expected percentages of different horn sizes which may be obtained. Then one requires to know when it may be necessary to reduce the female sector in order to maintain a balanced sex ratio. As yet we know nothing of the effects of unbalanced sex ratios — stochastic formulae tell us nothing of the behavioural implications.

All we do know is that most undisturbed wild populations approach closely to a 1:1 ratio; which tends to suggest that it is not advantageous for the survival of a species to depart widely from this ratio. Management for hunting, therefore, requires a knowledge of the numbers of animals in an area; their sex ratio; the recruitment rate; the rate of individual growth, and the ecological longevity. A knowledge of growth and age will tell us when trophy animals will become available in a population, and the ecological longevity may tell us how long, on average, a trophy may be available in the population.

Cropping, obtaining a sustained yield of meat or hides, is not a new concept in Africa: but its application in the context of studied game use is. It implies the taking of the maximum possible numbers of a species, a difficult task for the wildlife manager. If too few animals are taken the operation may be uneconomical, if too many are taken the resource may be destroyed. Even if the optimum number is taken the age structure of the population will be changed from a natural one to an artificial one. In American and European literature the survival curves given for large mammals (usually deer) are often concave. This is not the natural curve of survival that most wild, large mammal populations assume, as was once thought. The preponderance of young animals implied by the concave curve is the result of elimination of the adults by hunting. Most African ungulate populations that have been studied tend to show a convex curve which indicates a preponderance of adults with a low recruitment rate.

A large adult sector means, however, that there is ample potential for recruit production. What we have to find out for obtaining a maximum sustained yield, is the number which can be removed from the population so that the population will recruit at the maximum possible rate.

Several equations have been proposed for determining the maximum sustainable yield of a population, but all possess the shortcoming of a lack of predictability of response in real life. The actual numbers of removable animals in the equations can only be found by trial and error, although admittedly within defined limits but if we reduce our population the habitat may respond with a reduced carrying capacity; if, for example, the original numbers had been self-maintained by a high grazing pressure. Unpalatable grasses may be given a chance to invade. The density of our reduced population may also start to oscillate violently due to changes in predator–prey ratio, disease immunity ratios, or disruption of social organization; although this would only be identified as a decline below the numbers the population had been lowered to. Thus the management of populations for the economic production of hides or food, in its early stages such as we are concerned with in Africa, is a complex subject. Once populations have been reduced to an artificial status, management becomes more simplified in its understanding, but more intensive in its application. In other words, once you are keeping a population under control by artificial means, one must always apply that control. Otherwise large increases in the population may take place, followed by massive deaths when the carrying capacity of the habitat is exceeded.

THE ROLE OF RESEARCH

Russell made it quite clear in his work *Management policy in the Tanzania National Parks* published 1968 that research should be secondary to the primary role of a national park as a sanctuary. Thus, although today the scientific objectives have become those leading to management of an area in respect to maintaining its status as an ecological island, some parks link this to a preservation policy, while others link it to a policy of exploitation for public enjoyment. But sound research is the basis of objective management, and since national parks are unlikely to maintain their ecological integrity in isolation, placing research on a secondary footing can only imply *laissez-faire* attitudes which are ultimately detrimental to the animals and plants that have been given sanctuary. There seem to have been no conflicts concerning research outside of national parks, but little has been achieved in the way of management benefits. Most research, to date, in East Africa has consisted of cataloguing, whether it be cataloguing behaviour, ecology, physiology or anatomy. There has been almost no experimental manipulation, reducing or increasing species of plants or animals to observe the effects; with one notable exception, the reduction of the hippopotamus in parts of the Rwenzori National Park of Uganda. A great deal of background research has now been achieved in East Africa, but it is my fear that it is now going to stagnate along the lines outlined above, because of a lack of serious application of management.

CONCLUSION

We have the methods and we have the knowledge to cover many aspects of wildlife management; what has been noticeably lacking is confidence in applying these methods and knowledge to the management of African wildlife. This I attribute to two factors: lack of defined objectives, and lack of management plans to carry out these objectives. Whether it is a national parks organization or a game department organization, the first requirement is an overall management plan, clearly defining the department's objectives. Each individual area or national park should then have, before a single road or building is constructed, a management plan applying the objectives to that area. There should be no necessity for discussion as to whether a species should be culled or not, the situation should be statutorily defined in the master plan. I doubt whether a single national park or game reserve exists, which had a management plan drawn up before the construction of a road in the area. Wildlife management, as it is presently applied, exhibits a lack of professionalism. Let us hope that the new fields open to its application in western Africa will profit from the mistakes of others, and lead to enhanced practical achievements.

REFERENCE

PHILLIPS, J. C. (1935) *The London Convention for the Protection of African Flora and Fauna.* (Special Publication of the American Committee for International Wildlife Protection, No. 6.)

23

MANAGEMENT PLANNING AND PRACTICE: CASE HISTORIES FROM MALI AND BENIN

J. A. Sayer

F.A.O., B.P. 506, Cotonou, Republic of Benin

INTRODUCTION

A large proportion of West African national parks lie in a narrow band close to the 1000 mm isohyet in areas where poor soils and human and animal diseases prevent extensive use by man (Figure 1). Many of these parks have similar management problems. I would like to discuss them using examples from two areas: the Baoulé National Park and its surrounding reserves in Mali and the Pendjari and W. du Niger National Parks in the republic of Benin.

All three parks lie in the Sudan or Sudano-Guinean savannah zone. Over large areas the soils are shallow and the surface is covered with a laterite crust.

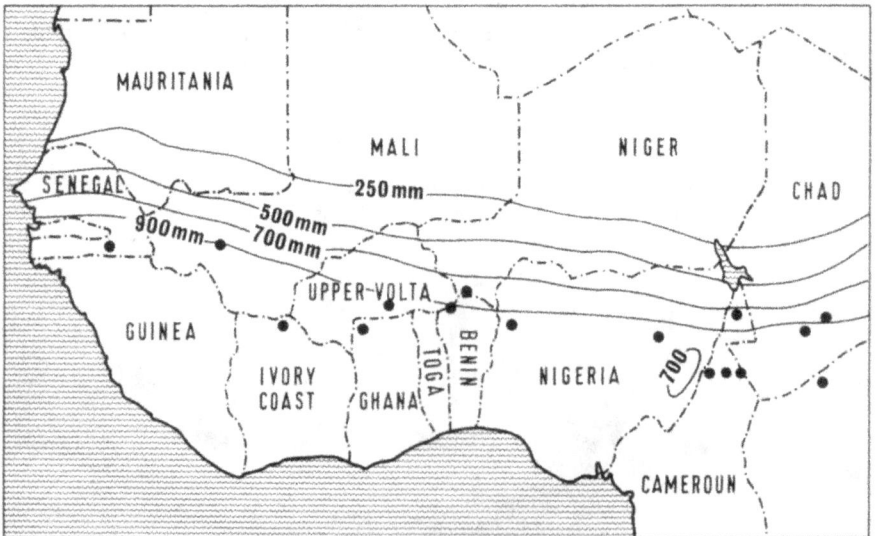

Fig. 1. Distribution of National Parks in West Africa in relation to rainfall (data from Michelin Map 153 of Africa North and West, 1973).

194

These areas support a dense scrubby vegetation which burns annually; the commonest trees are species of *Combretum* and the grasses species of *Loudetia* and *Andropogon*. On lower areas where soil accumulates there is a tall grass savannah where the main trees are *Terminalia* spp., *Isoberlinia doka* and *Butyrospermum parkii*. Fadamas and lagoons occur in all the areas but are particularly extensive in the Pendjari, where they constitute key areas for wildlife and a major attraction for visitors. Fringing riparian forest occurs along the major rivers but in the Baoulé and the W. du Niger it has been damaged by cutting and fire.

The large mammals of the areas are similar (Table 1) except that giraffe *Giraffa camelopardalis* and giant eland *Taurotragus derbianus* occur only in the Baoulé, and Topi *Damaliscus korrigum* are only readily seen in the Pendjari although they occur in the W. du Niger in small numbers, and were recorded

TABLE 1

Distribution of large mammals in the Benin National Parks

	Baoulè	Pendjari	W. du Niger
Papio anubis, Anubis baboon, Babouin doguera	x	x	x
Cercopithecus aethiops, Vervet monkey, Vervet	x	x	x
Erythrocebus patas, patas, patas	x	x	x
Lycaon pictus, wild dog, cynhyène	?	?	?
Crocuta crocuta, spotted hyaena, Hyène tachetée	0	0	0
Panthera leo, Lion, Lion	0	x	0
Panthera pardus, Leopard, Léopard	0	0	0
Acinonyx jubatas, Cheetah, Guépard	?	0	0
Loxodonta africana, Elephant, Eléphant	0	0	x
Hippopotamus amphibius, Hippopotamus, Hippopotame	x	x	0
Phacochoerus aethiopicus, Wart-hog, Phacochère	x	x	x
Giraffa camelopardalis, Giraffe, Girafe	0	–	–
Taurotragus derbianus, Giant eland, Elan de Derby	0	–	–
Tragelaphus scriptus, Bushbuck, Guib harnaché	x	x	x
Hippotragus equinus, Roan antelope, Hippotrague	x	x	x
Kobvs defassa, Defassa water-buck, Cobe defassa	x	x	x
Kobus kob, Kob, Cobe de buffon	–	x	x
Redunca redunca, Reedbuck, Redunca	0	0	0
Alcephalus buselaphus, Western hartebeest, Bubale	x	x	x
Damaliscus korrigum, Topi, Damalisque	?	x	0
Gazella rufifrons, Red-fronted gazelle, Gazelle à front roux	?	–	?
Cephalophus rafilatus, Red-flanked duiker, Céphalophe à flancs roux	x	x	x
Sylvicapra grimmia, Grimm's duiker, Céphalophe de Grimm	x	x	x
Ourebia ourebia, Oribi, Ourébi	x	x	x
Syncerus caffer, African buffalo, Buffle d'Afrique	0	x	x

x: readily seen by visitors.
0: present but seldom seen by visitors.
?: status not known.
–: absent.

from the Baulé in the 1950s. Visitors see a larger number and variety of animals in the Pendjari than in the other two parks, lions *Panthera leo* are seen by a large proportion of visitors. Elephants *Loxodonta africana* are present in all three parks but rarely seen; they are most numerous in the W. du Niger.

The only large mammals that might be considered to be immediately endangered in any of the Parks are giraffe, buffalo and eland in the Baoulé, where they are intensively hunted around the water-holes, which they use in the late dry season. Buffalo and eland are commoner in forest reserves in the Mandingue mountains, south of the park. The wild dog *Lycaon pictus* is now absent or very rare in all three areas although it was frequent in the recent past, they were often shot by national park's staff as they were considered pests but the recent decline is more probably attributable to disease. Wild dogs are susceptible to canine distemper and it is believed that this has caused their decline throughout Africa.

The Baoulé and W. du Niger National Parks were created in 1954; both had already had game reserve status for several years. The Pendjari was created as a protected forest in 1954, a total reserve in 1955 and a national park in 1961. Application of the laws has been concentrated in areas which were exploited for tourism; thus the Baoulé National Park has received very little management while the Fina reserve has been treated as a national park and exploited for tourism. Until recently visitor hunting in the Baoulé and Pendjari hunting zones, near these respective National Parks, has been more developed than non-hunting tourism within the parks. An anomalous situation developed whereby animals were more readily visible in the hunting zones than in the parks; this was presumably because the activities of subsistence hunters were more effectively suppressed by the presence of licenced hunters accompanied by guards in the hunting zones than they were in the parks which were seldom visited. This situation did not develop in the Pendjari since, although the Dahomean hotel was only built in 1967/69, visitors had previously entered the park from the nearby hotel at Arly in Upper Volta.

The Park, reserves and hunting zones of the Baoulé area, comprise almost one million ha with no permanent human settlements. The Pendjari and the W. du Niger lie in a belt of land stretching from the Niger river south of Niamey to the border of Togo in which there are very few settlements in about two million ha. The areas are probably depopulated because of onchocerciasis (river blindness) and trypanosomiasis (sleeping sickness); all three have been densely populated in the past. Apart from their low human population the parks are typical examples of the vegetation belts in which they lie. An unusual environmental feature of the Baoulé is a broad alluvial belt which is a relic of the geological period when the upper Niger drained through the Baoulé into the Senegal river; the Baoulé area also contains examples of the sandstone inselberg landscapes of the Mandingue plateau. The Pendjari Park lies in the Volta depression where there are a high proportion of seasonally waterlogged soils. These areas support open grasslands or parklands, landscapes which are attractive to visitors and where animals are easily seen. The quartzite cliffs of the Atacora mountains form an interesting landscape feature of this park.

CONSERVATION MANAGEMENT

Although the present human populations of all three areas are low, the ecology of all three has been greatly modified by man's past activities. Figure 2 shows the rate of population increase in West Africa in recent centuries. The rate of increase before 1900 was slow, furthermore most of this increase took place in the wetter coastal regions where the staple food crops, maize, cassava and yams were introduced from South America in the fifteenth century. The livestock and cultivated millet and sorghum of the Sudan zone have been farmed for over 2000 years and it was in the Sudan zone that the main centre of human population was found in the past.

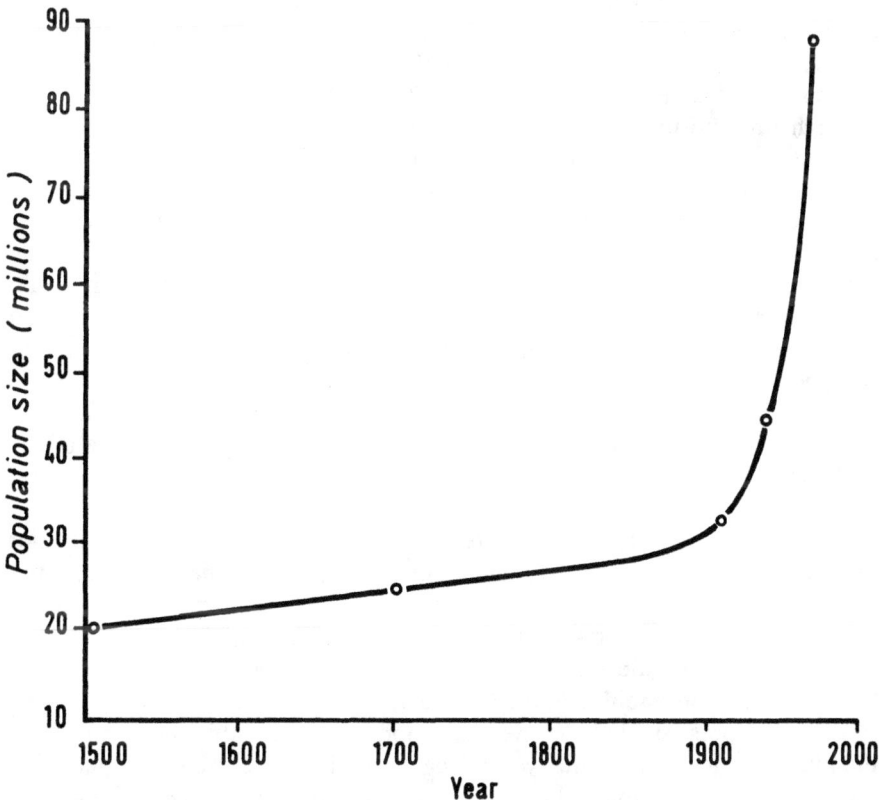

Fig. 2. Approximate pattern of human population growth in West Africa showing the relatively slow rate of increase before the twentieth century.

Two of the most important medieval states in West Africa were those of Ghana and Mali, which dominated the region from the fifth to the fifteenth century. The Baoulé area lies near the centre of these ancient empires and its ecology must have been affected by the high population densities that occurred at that time. From the seventeenth to the nineteenth century the smaller Kaarta kindgom had its capital near the Baoulé river only a few kilometres from the present park boundary.

The human population of the Baoulé Park is probably lower now than at any time in the past 1500 years, and present conditions might therefore be exceptionally favourable for large mammals.

The history of the Pendjari and W. du Niger Parks is less well known but the presence of human artifacts and atypical vegetation indicates protracted and extensive human use. The Pendjari was, in the past, an important area for iron smelting (Oliver, 1967).

Hunter (1966) has suggested that onchocerciasis causes cycles in the human populations of riverine areas in the Sudan zone, the wavelength determined by the period during which the disease built up to a level where its debilitating effects caused people to emigrate or die. If this were correct low population levels may have recurred periodically in the past.

Man has changed the ecology of West African savannahs, principally through fire and the grazing of his domestic animals. Fire is considered by Bartlett (1956) to have been an important influence on West African vegetation in pre-Christian times, and cattle have been present for 'thousands of years' (Deshler, 1963), although it was not until the eighth century that zebu cattle arrived from North Africa or the Middle East (Fage, 1969). Aubreville (1949) considers that the natural vegetation of West African savannahs would be a dry forest, however, since the present climatic conditions have only existed for about 5000 years 1969) true climax vegetation may not have had time to develop.

The present flora and fauna of the savannah zone national parks are not natural and the large mammals which constitute an important feature of the parks for visitors would not occur in such variety or abundance in natural vegetation conditions.

The ecological changes which are taking place in the parks now do not necessarily reflect present management and some of them may be the continuation of long-term changes initiated when human populations were higher. This is almost certainly the case of the extensive gully erosion, which occurs along the Baoulé and Mekrou rivers. A large amount of wood must have been used for iron smelting in the Pendjari, the iron ore appears to have been transported, presumably to the nearest wood supply, and this may account for the impoverished state of the woody vegetation in the centre of the park. All three parks contain examples of *Butyrospermum paradoxum* savannah, a vegetation type produced by cultivation of *Isoberlinia doka* woodland.

An attempt was made in the Baoulé to correlate the grazing of domestic stock with habitat conditions and wildlife use. The intensity of use by wild and domestic animals was estimated from dropping counts in 400 m² strip samples. The number of trees of different size classes, the perenial grass cover and bare ground were recorded. The habitat variables were influenced by catenary position and latitude; in general the changes that occurred towards the top of the catena paralleled those associated with increased latitude. The proportion of bare ground increased with latitude, the perennial grass cover decreased, the density of trees over 6 m high decreased and those less than 6 m increased.

Table 2 gives some examples of the results of this work. The regression equations for perennial grass cover, tree density and the percentage of bare ground, give a measure of the condition of the habitat at a point in time. Comparisons could be made with other physically similar areas or with the same area after a time interval. The shift of the regression lines for bare soil and perennial grass against minutes of latitude could be considered as a quantitative measure of the annual rate of advance of desert conditions.

The data were not extensive enough to permit good comparisons to be made between the distributions of wild and domestic animals. My impression from travelling through the park was that wart-hog *Phacochoerus aethiopicum* and oribi *Ourebia ourebia* were more numerous in areas where cattle occurred and that the larger antelopes were less numerous. The data suggest that there is no correlation between the distribution of cattle and that of wildlife. The apparent frequency of smaller wild animals in areas where cattle occur is probably the result of greater visibility in areas subject to cattle grazing.

In the north of the Baoulé Park, bush encroachment is occurring as a result of overgrazing. Most of the common woody species are not browsed (Table 3), three rarer species which are heavily browsed are *Acacia seyal, Boscia angustifolia* and *Opilia celtidifolia*. In the Baoulé it appears that an increase in browsing animals would not be sufficient to reverse the bush encroachment but in the Pendjari, where animal densities are higher, the range of woody species used is much higher.

Conservation management policy should aim to keep human influence to a minimum. Fire would be excluded completely if it were practicable to do so. Early burning has been adopted as an expedient, as it is thought that this constitutes the lesser of two evils. In reality even extensive early burning is difficult to achieve and large areas burn late in the dry season.

Attempts have been made to prevent all hunting in the parks. Anti-poaching activities occupy a large proportion of staff time, poaching is kept at reasonable levels in areas used by tourists, elsewhere it is excessive.

There is no illegal grazing of domestic animals in the Pendjari, the problem exists in the W. du Niger where it has not yet been studied, in the Baoulé the problem is severe. Large numbers of cattle, sheep and goats from the Sahel concentrate along the northern Baoulé in the dry season: it has proved difficult to enforce anti-grazing laws, particularly during the recent drought when the Baoulé was the only source of water.

TABLE 2

Examples of correlations and regression equations for some environmental parameters in the Baoulé National Park

x	y	Correlation coefficient	Regression equation
Minutes of latitude North of 12° N (inversely proportional to rainfall)	% Basal cover of perennial grasses	$c = -0.276$ p at $0.02 = 0.274$	$x = 4.21y - 0.34$
"	% Bare soil	$c = +0.279$ p at $0.1 = 0.275$	$x = 3.01y - 0.54$
"	Density of trees N° per 400 m²	$c = 0.135$ p at $0.1 = 0.231$	$x = 10.6y + 0.37$
% Basal cover of perennial grasses	Density of trees less than 3 m	$c = -0.247$ p at $0.1 = 0.231$	$x = 5.33y - 0.51$
% Basal cover of perennial grasses	Density of trees greater than 6 m high	$c = 0.120$ p at $0.1 = 0.231$	$x = 2.34y + 0.14$
Frequency of cattle droppings	% Bare ground	$c = 0.455$ p at $0.1 = 0.476$	$x = 96.9y + 0.08$
Frequency of droppings of all wild ungulates	Frequency of droppings of domestic ungulates	$c = -0.205$ p at $0.1 = 0.275$	$x = 3.01y - 0.54$
Frequency of droppings of cattle	Frequency of droppings of large antelopes	$c = -0.096$ p at $0.1 = 0.476$	$x = 1.27y - 0.42$
" " "	Frequency of droppings of wart-hog	$c = -0.223$ p at $0.1 = 0.360$	$x = 1.91y + 2.30$

TABLE 3
Intensity of late dry season browsing in Baoulé National Park

Species	No of individuals in sample	% browsed
Acacia macrothyrsa	1	—
Acacia seyal	23	100
Acacia sieberiana	3	—
Adansonia digitata	1	—
Anogeissius leiocarpus	14	—
Bauhinia reticulata	6	—
Bauhinia rufescens	2	—
Bombax costatum	29	—
Borassus aethiopium	1	—
Boscia angustifolia	2	100
Burkea africana	4	—
Butyrospermum parkei	10	—
Carapa prosera	5	—
Cassia sieberiana	1	—
Combretum elioti	1	—
Combretum ghasalense	125	—
Combretum micranthum	124	—
Combretum nigricans	3	—
Cordia mixa	3	—
Cordyla pinnata	3	—
Dichrostachys glomeratus	6	—
Diospyros mespiliformis	16	—
Entada africana	7	—
Fagera xanthoscyloides	11	—
Fuiroutia canthoides	1	—
Gardenia	—	—
Gardenia ternifolia	11	—
Gardenia senegalensis	7	—
Guirea senegalensis	6	—
Isoberlinia doka	8	25
Landolphia senegalensis	2	—
Lannia acida	3	—
Mitragyna inermis	4	—
Oxytenonthera abyssinica	35	—
Pourparea burea	1	—
Pterocarpus erinaceus	15	—
Pterocarpus luscens	11	—
Spongia	1	—
Sterculia sitigera	3	—
Tamarindus indica	7	—
Terminalia avis-senalis	1	—
Terminalia macroptera	11	—
Ximenia americana	9	—
Ziziphus jujuba	3	—
Ziziphus mauritianus	4	—
Tree spp. misc.	52	6.25

MANAGEMENT PLANNING AND PRACTICE

Management plans have been based on a proforma given in the *American Wildlife Society Handbook* (published in 1971); their main function is to make all basic information about the area readily available; detailed prescriptive planning has not been included. The term 'master plan' as used by the American National Park Service would be more appropriate. It has been found

TABLE 4

Proforma national park master plan

1. *Introduction*
1.1. General information
1.2. Major surveys and maps. Includes a list of major studies, of all maps and aerial photographs and of important legislation
2. *Reasons for establishment*
2.1. Primary interest
2.2. Secondary interest
2.3. Value
3. *Description of the area (maps used wherever appropriate)*
3.1. Geographical situation
3.2. Climate. Summaries of all available data
3.3. Geology and geomorphology
3.4. Hydrology
3.5. Prehistory and ecological history
3.6. Land-use history
3.7. Vegetation. Depending on the depth of the study this might include a general description of the vegetation, a vegetation map or ideally a landscape classification with information on vegetation in the various land systems or land facets
4. *Wildlife population status*
4.1. Introduction. Mention wildlife of outstanding interest
4.2. Detailed information for different groups of animals. Check lists appended
5. *Needs and activities of human populations in the area*
5.1. Detailed information on different types of activity and different ethnic groups in the area and on potential users of the park

PART 2 PRESCRIPTIVE

6. *Estate Management*
6.1. Roads, fences, airstrips, maintenance of buildings, plant and equipment. Management necessary to facilitate visitor use
6.2. Interpretation program. Signposting, printed information, stopping points, reception centres
7. *Conservation management*
7.1. Law enforcement
7.2. Fire control
7.3. Fishing
7.4. Wildlife management
8. *Inventories and research*
8.1. Needs for continuous monitoring of habitat changes and status of wildlife

APPENDICES

1. Bibliography. A complete bibliography is desirable, not just a list of references cited.
2. Annotated check lists of fauna.

useful to collect data in a standardized way (Table 4), index cards have been used for all species of flora and fauna and for bibliographical references. Copies of all available maps, aerial photographs and the relevant legislation have been included. A chronological record of historical events concerning the park and its region has proved useful. Information about individual sites in a park can be collected using numbered references to a sketch map; the map may be incorporated into the plan. Such a map can show the location of features of faunal or vegetational interest, historical sites, outstanding landscapes and areas where management is required.

The only management activities in the parks are the enforcement of laws against poaching and grazing, the control of fire, the maintenance of roads and the control of visitors.

The Pendjari Park has two good hotels both air conditioned, one in the centre of the park and the other in the nearby hunting zone. There are also three small camps run by professional hunters whose clients visit the park. There are 200 km of seasonal roads inside the park and a further 180 km in the hunting zones, all managed by forestry staff. Law enforcement staff are stationed in peripheral villages and at the hotel in the park: during the dry season, they patrol on mopeds. Seasonal road maintenance and early burning is carried out by these staff; a truck, tractor and bush cutter are used. A number of additional roads are being constructed to provide variety for visitors, to lessen pressure on existing roads and to facilitate law enforcement. An interpretation programme has been established to provide signposts, maps, booklets and observation points. The park is administered from Natitingou, 110 km from the park entrance, on a moderately good road.

The W. du Niger has no tourist accommodation in Benin but camping is allowed at a waterfall on the Mekrou river and there is a rest house at Banikoara, 14 km from the park entrance. There is also a camp run by a professional hunter on the park boundary. There are 47 km of road in the part of the park in Benin, a further 100 km in Upper Volta (where there are no visitor facilities) and a network of tracks in Niger where there is a hotel on the park boundary. Law enforcement staff are stationed in villages around the park, they patrol by moped and are administered from Kandi, 110 km from the park entrance on a good road. A road is planned to connect the Benin part of the park with the Niger part with loop roads to sites of special interest. Deposits of phosphate exist in the park which will probably be exploited commercially in 15 to 20 years, when the necessary infrastructure has been established. Extraction will necessitate a railway or major road through the park and a resident staff of several hundred people and their dependents.

The Baoulé has three small self-catering camps which were originally established for hunters; from these camps, access is possible into the Fina reserve where a good variety of animals can be seen: there is no access into the National Park. There are a total of about 350 km of road in the protected areas which are maintained by the Forestry Department. Law enforcement staff are stationed in villages fringing the protected area. Staff are supervised from Bamako but access is only possible in the dry season.

A major factor limiting the development of the parks is that the senior staff responsible for their administration are based in towns a long way from their management staff. This increases transport costs and lessens contact between senior staff and the day-to-day activities of park staff. The situation at La Tapoa in the W. du Niger is preferable where the park is managed from a centre, which is on the park boundary; the single park vehicle is based there and is thus constantly available for use in the park. The West African practice of placing law enforcement staff in villages, contrasts with the East African system where they are usually stationed in the area which they are supposed to protect.

The conflict between nature conservation and economic exploitation presents a fundamental problem to the managers of the parks. All were established under statutes which clearly state that their primary role is to protect the natural environment, but there is a tendency for the agencies responsible to consider that commercial tourism is the real objective. This is understandable in poor countries but it seems a pity that overseas aid which comes mainly from the U.N. agencies is generally available only for nature conservation projects which are considered profitable.

REFERENCES

AUBREVILLE, A. (1949) *Climats, Forêts et Désertification de l'Afrique Tropicale.* (Société d'éditions Géographiques, Maritimes et Coloniales, Paris.)

BARTLETT, H. H. (1956) Primitive Agriculture and Grazing in the Tropics. In *Man's role in changing the face of the earth.* Thomas, W. L. (Ed.). (Univeristy of Chicago Press.)

DESHLER, W. (1963) Cattle in Africa: Distribution types and Problems. *Geog. Rev.,* Vol. 53, pp. 52–8.

FAGE, J. D. (1969) *A History of West Africa.* (Cambridge University Press.)

HUNTER, J. M. (1966) River Blindness in Nangodi, Northern Ghana: A hypothesis of cyclical advance and retreat. *Geog. Rev.,* Vol. 56, pp. 398–416.

OLIVER, (1967) *West Africa before the Europeans.* (Methuen, London.)

24

AN ECOLOGICAL MANAGEMENT PLAN FOR THE KAINJI LAKE NATIONAL PARK

S. S. Ajayi and J. B. Hall

Department of Forest Resource Management,
University of Ibadan, Nigeria

INTRODUCTION

In Nigeria, the role of game reserves in conserving wildlife for various purposes (but particularly as a tourist attraction) is widely recognized. It is recognized also that the fauna constitute only one element of the complex ecosystems to which they belong and that these ecosystems are not necessarily in a stable state. Accordingly, for areas set aside as game reserves it is desirable that an understanding of the ecosystems present be developed so that any operation carried out is directed towards ensuring the establishment and maintenance of a situation which meets as effectively as possible the long-term requirements of management.

As stressed by Eggeling (1964) the same general principles for ensuring effective management apply for a variety of forms of land use including management of land as game reserves, there is a definite need for a formal management plan to be drawn. Such a plan has recently been prepared for Nigeria's Kainji Lake National Park (Ajayi and Hall, 1975). It is considered that a summary of the plan itself, will be of interest to those concerned with game reserve management elsewhere in West Africa.

The principal purpose of the plan is to provide, in an easily consulted form, all the available information relevant to the management of the reserve. In addition, the plan makes provision for its regular revision and updating and incorporates a timetable for these processes.

KAINJI LAKE NATIONAL PARK MANAGEMENT PLAN

The objective of management

It was considered that by making the area a game reserve the wildlife populations there could become a major tourist attraction. Events since establishment have demonstrated that this aim is realistic. Management continues to be directed towards creating an important tourist amenity.

State of knowledge — the ecological background

During compilation of the first management plan it became evident that a considerable body of information about the reserve existed, though in widely scattered reports and publications. No major aspect had been totally ignored and the relevance for management of much of this information was immediately apparent. In the brief summary given here, the essential features of the conditions in the reserve are considered under the headings:

- (a) General nature
- (b) Physical features
- (c) Climate
- (d) Vegetation
- (e) Human influence
- (f) Fauna

(a) General nature

The reserve extends about 80 km in an east–west direction and about 60 km north–south, its centre being located close to the intersection of latitude 10°N. and longitude 4°E. It occupies an area of approximately 4000 sq. km.

(b) Physical features

Physically the area displays considerable uniformity. The terrain is gently undulating but there is a general increase in elevation from Lake Kainji shore in the east (142 m) ridges and hills occur; the highest is 374 m. Because of the slope, the drainage is to the east, into the River Niger and Lake Kainji. From four small basins water flows into the lake but the main part of the drainage (from 60 per cent of the reserve) is into the River Niger via the River Olli.

There is little geological variation, all the surface rocks being referable to the Basement Complex. Local occurrences of schists, quartzite and granite are reflected in differing land forms and also, apparently, differing vegetation cover. Soils have been studied little but appear to be mostly shallow and of low fertility.

(c) Climate

Whilst climatic data from the reserve itself are at present insufficient to justify analysis, there is a very substantial volume of data from stations around it. Reference to this information not only indicates the general nature of the prevailing climate but also trends across the area and how, through effects on the water available for plant growth, this influences the availability of grazing and browse.

The climate is strongly seasonal, most factors studied being closely correlated in the trends shown. The rainy season lasts around 175–190 days during which time most of the 1050–1250 mm of rain falls.

In the rainy season, monthly means of daily temperature maxima are around

30°C; corresponding minima are 7–8°C lower. Monthly means of daily relative humidity measured at 09.00 hrs in the rainy season (roughly 2½ hrs after sunrise) from Yelwa are in the order of 60–80 per cent. In the dry season, the monthly means for daily temperature maxima rise to 35–40°C but the minima fall to only 14–15°C in the Harmattan and the resultant effects on animal movement have been noted. Relative humidities at 09.00 hrs (2 hrs after sunrise) are then barely 30 per cent at Yelwa and probably even lower in the reserve.

The variation in climate across the area is marked and affects both the recorded values for individual factors and the seasonality. Two trends seem superimposed over the area, which becomes increasingly wetter from north to south and from east to west. At the wetter extreme, mean annual rainfall is believed to approach 1250 mm, 200 mm more than in the drier parts. In addition, the rainy season starts earlier (mid April) in the wetter part but about two weeks later in the drier part. Throughout the reserve, however, the rainy season ends at the same time (mid October). An increasingly harsh temperature regime with higher maxima and lower minima accompanies the decrease in rainfall across the area.

From figures of potential evapotranspiration and estimates of the variables used to calculate these for stations around the reserve the commencement and length of the growing season have been estimated to be roughly 10 April and 220 days respectively in the wetter western section, roughly 25 April and 210 days in the drier eastern section and 5 May and 200 days in the north.

The levels of net radiation favour the predominance of high photosynthetic capacity plants in the sward. Though these are noted (Black, 1971) to be efficient in utilizing available water, their nutritional values per unit weight are low (Janzen, 1975).

(d) Vegetation

The vegetation is the most thoroughly studied aspect of the reserve environment. Several reports on the vegetation have been published, the most recent (Geerling, 1975) providing the main basis on which the reserve is currently stratified for management.

In addition to riverine vegetation complexes and other communities which are distinctive but occur as isolated units small in extent individually, Geerling has recognized five major types of upland vegetation. Together, these five types cover over 70 per cent of the reserve. They are:

A. *Acacia* Savannah woodland
B. *Burkea africana/Terminalia avicennioides* Savannah woodland (typical)
C. *B. africana/T. avicennioides* (*Acacia* variant)
D. *B. africana/T. avicennioides* (*Afzelia* Savannah woodland variant)
E. *B. africana/T. avicennioides* (*Detarium* Savannah woodland variant)

These major types are not sharply differentiated and grade into one another. Geerling's decision to recognize these units, though arbitrary, was aimed at a practical subdivision of the range of variation represented in the savannah.

Interpretation of the vegetation in environmental terms emphasizes that this was a constructive approach. Correlation with the underlying geology is reflected in the limitation of the *Acacia* savannah woodland to the area of schists near the lake and of the *Acacia* variant (intermediate between the former and the typical *B. africana/T. avicennioides* savannah woodland) to the schist/gneiss transitional zone west of the schists. The *Detarium* savannah woodland variant also displays a relationship with rock type, being associated with the presence of plinthite. Typical *B. africana/T. avicennioides* savannah woodland occurs in most parts underlain by gneissose rocks but grades into the *Afzelia* savannah woodland variant in the parts subject to heavy rainfall where disturbance seems to have been least.

(e) Human influences

Human population density has been very low for a long time and evidence within the reserve of past human activity is negligible, suggesting that the present vegetation is in a fairly stable state. Water shortages and the tsetse problem have been considered probable reasons for the low population density (Howell, 1968).

Fire occurs annually, affecting most of the area. Recently, controlled early burns have been used to improve visibility, but extensive late burning still occurs, mainly in conjunction with poaching.

(f) Fauna

The fauna is basically that typical of West African Guinea or Sudan savannah with some additions reflecting special habitats present (e.g. manatee in Lake Kainji). Various check lists have been prepared, including ones for mammals (Child, 1974), birds (Afolayan, 1973), reptiles (Child, 1974) and fish (Ita, 1972; Child, 1974).

Achievements to date

Management to date relates to two broad areas. The Kwara State Forestry Service has been principally concerned with the organization of the reserve in terms of improving access, with developing a staff to run the reserve and with combatting poaching and other illegal activities within the reserve. These efforts are continuing but already it has become possible to open the reserve to tourists and to guide them to points where the prospects of seeing game are good. Pelinck (personal communication) reports that visitors are generally well satisfied with what they see.

Working in close co-operation with the State Forestry Service and with individuals in institutions elsewhere, especially Universities, staff of the Kainji Lake Research Institute (formerly the Kainji Lake Research Project) have undertaken a considerable amount of research work in the area. Much of this has, of necessity, been concerned with reconnaissance surveys which have provided information that is now being used to stratify the reserve for management purposes and with the preparation of check lists of various groups of organisms. Whilst this survey aspect of research still continues, the information so far collected has enabled much more ambitious studies concerning ecosystem dynamics to be initiated and it is the results of these that are expected to indicate the most promising lines of practical management for the future.

Much attention has been devoted to the large mammals. Rough estimates of population density have been made at intervals in order that any fluctuations in numbers may be detected. Information on herd structure and on diurnal and seasonal patterns of behaviour has been collected and animal activities in relation to vegetation type, and in some cases burned areas also, have been recorded.

The major role of the reserve as a tourist attraction has been appreciated from the outset and surveys of the opinions of visitors on their visits have been made so that these may be taken into consideration as new developments are undertaken. In addition, the problem posed by differential visibility for viewing game — on a seasonal basis and in different vegetation types within the reserve — has been investigated and the results taken into account in recent management practice.

Future investigations

Most of the research investigations initiated so far have already yielded information which has proved useful for formulating management policies. Nevertheless, it is appreciated that because there are marked year-to-year climatic variations, study of many aspects of the reserve's ecology must be continued for several more years.

The maintenance and pursuance of the investigations already initiated is essential if the practices adopted for management are to be continually refined by the incorporation of modifications based on improved understanding of the reserve's ecology. This is work of high priority.

Two other areas of investigation also demand high priority and need to receive attention soon. As a sociological study, the poaching community should be examined. Until the outlook and methods of operation of the poachers is properly understood it will be impossible to control poaching efficiently. Secondly, with a view to assembling data permitting carrying capacities to be estimated for the reserve, special attention must be paid to a detailed survey of the sward composition throughout the reserve and to the estimation of vegetation biomass and its seasonal changes. The information obtained should be related to the activities of individual species by faecal analyses and leaf epidemics studies. In conjunction with a sward composition study, an appraisal of the condition of the grassland and how this may be changing, together with a more detailed examination of the associated soils, would be valuable.

It is obvious that at least in the foreseeable future fire will be the most powerful tool available for management and assessment of the long-term effects of the burning/grazing treatments being applied to the *Isoberlinia* woodlands, the *Terminalia macroptera* unit of the River Olli complex and the *Detarium* savannah woodland is desirable.

In addition, as staff and funds permit, similar investigations in other vegetation types — especially typical *Burkea africana/Terminalia avicennioides* savannah woodland and its *Afzelia* savannah woodland variant — should be commenced.

All the areas of investigation indicated fit together well in the general study of the reserve's ecology but more information is needed on several specific problems such as the failure of the valuable browse species *Afzelia africana* to regenerate and the apparently unsatisfactory status of animals such as the crocodile.

The major gap in the understanding of the reserve's ecology is the absence of any information on the insect populations. Efforts are therefore being made to involve suitably qualified people in research into this aspect as a matter of urgency since it is clear from studies elsewhere in West Africa (Lamotte, 1975) that it is of the greatest importance in savannah ecosystems.

Present and future administration and management

The two aspects of the Game Preservation Unit's structure which at present are most in need of periodic review are the staff complement and its mobility. The existing staff is relatively small to operate effectively over such an extensive area. The most senior of the staff is a graduate with the status of Senior Assistant Conservator of Forests (equivalent to Senior Wildlife Officer). His immediate subordinates are four Game Rangers, men educated to Secondary School level (West African School Certificate) and the remainder of the 40 man total complement are Game Guards, who require no educational qualification to be recruited.

As circumstances permit, members of this staff are sent for further training of a more specialized nature at appropriate levels and of varying duration but on-the-spot training has been used in most cases.

A staff strength of around 130 is needed to ensure that the development needed over the next few years is carried out. The most Senior Officer should be Principal Game Warden and his deputy should be a Senior Game Warden under whose direction would be three Game Wardens responsible for the three fields: patrol and enforcement, park management, construction and maintenance. All these five officers should be of graduate or equivalent status. The Game Ranger strength should rise to 10 and for each Ranger there should be eight Game Guards. The establishment would be completed by appropriate service staff-drivers, mechanics, works inspectors, etc. It is likely that facilities for short- and long-term specialist training in wildlife and conservation for all levels of staff will be available within Nigeria shortly and full advantage should be taken of these for in-service training of staff.

To mobilize this expanded staff a much larger transport fleet will be needed, specialized construction vehicles will be essential if the road network is to be improved and maintained. In July 1975 there were eight road vehicles and a single boat. Of the former, four were tractors, one a lorry and the others four-wheel drive cross-country vehicles. The needs of the projected construction programme alone are five tractors, three tippers, two bulldozers, two graders, two rollers and two small four-wheel drive vehicles such as land rovers. The headquarters and anti-poaching needs are mostly for small vehicles (seven suitable for cross-country work and five for carrying personnel along the better quality tracks) but lorries (two) are needed and four boats for work on the lake. There should be a radio facility for all vehicles on anti-poaching duty to maintain contact with headquarters.

On the organizational side of the reserve management, areas of priority for further development relate to poaching control, visitors' access and the location of the base camp and the tourist lodge. The area involved is large and effective anti-poaching work requires both improvement and extension of the existing network of roads so that all parts of the reserve can be reached by patrols with minimal delays. A boundary track is needed and although this is a major undertaking which may require some years for completion it should be commenced without delay. From the main game viewing area, roads along watersheds should enable any part of the reserve boundary to be reached easily by the Game Wardens and their staff as events require.

Additional roads for visitors are needed to permit access to the lake shore section of the reserve from Wawa and from the River Olli. It may be appropriate to combine the construction of such roads with re-siting the tourist lodge at a point where it can expand without unfavourable effects upon the reserve. In connexion with this need a new site on the lake shore just outside the reserve boundary has been proposed. A re-siting of the base camp to the reserve boundary is equally desirable and in this case an additional reason is to escape the water-supply problem which affects at the existing location.

DISCUSSION

It is clearly relevant to draw attention to the major problems arising with game reserve management in West Africa, although there is no special provision for detailing these in the management plan. It is useful to be aware of how West African conditions compare with those elsewhere in the continent where reserve management is concerned. It should be appreciated that it is the nature of the problems faced which strongly influences the priorities attached to different research and management activities. There are three main problems faced at present — the high incidence of poaching, the inadequate understanding of the ecosystems in the reserve and the small amount of previous work in comparable situations elsewhere in Africa.

The poaching problem is serious. Because the reserve is so extensive and the reserve staff is too small and because of very serious difficulties with fuel and transport in recent years, the incidence of poaching has not been significantly reduced and may even have increased. If poaching pressures continue rising unchecked the future of the area as a game reserve will come into question. An anti-poaching policy advocated is the involvement of the hunters concerned in more permanent positions in the game reserve establishment. This is not readily achieved at Borgu, however, as the major part of the hunting appears to be by parties whose home villages are hundreds of kilometres away. Alternative solutions are needed and to formulate these more must be understood of the hunters themselves.

The need for a more detailed understanding of the reserve's ecosystems arises from the desirability of maintaining these in a relatively natural state. In much forestry work, stress is placed on manipulating an ecosystem to direct as large a proportion as possible of the cycling nutrients into a harvestable product — generally cellulose. In the Borgu Game Reserve situation it is more important to first develop a knowledge of the ecosystems and the processes taking place in them and associated with the game populations. The intention is to predict how changes in management are likely to affect these so that any interference will be directed towards the upgrading of deteriorating site conditions or to the maintenance of certain conditions.

Because of the contrast in conditions between West African Guinea Savannah and areas set aside for wildlife conservation elsewhere, management techniques need to be developed locally. Where game reserve management is most highly developed, the vegetation is physiognomically simpler, with a sparse woody cover. While fire remains the major management tool, its effects are less strongly influenced by the timing of burns. In West Africa, the burning regime determines the vegetation structure very closely — there is a high annual production of combustible matter and grassland or scrub may be readily encouraged as desired. There have been several recommendations that appropriate burning policies be employed to thin out the woody plants and increase visibility. However, careful research into the balance between burning, visibility and soil erosion is needed before distinctive policies on burning can be formulated, in view of the five month period each year (at Kainji Lake) when rainfall is heavy. Thus, great importance is attached to investigation of the effects of different burning regimes on vegetation structure and site stability and to interactions of these with other ecosystem components. For most of the West African savannah such an emphasis would be equally relevant.

ACKNOWLEDGEMENTS

We are grateful to Messrs. T. Afolayan, J. Ayeni, C. Geerling and K. Milligan for useful discussions on the Kainji Lake National Park and constructive criticism of the first draft.

REFERENCES

AJAYI, S. S. and HALL, J. B. (1975) A management plan for Borgu Game Reserve. (Unpublished MS Dept. of Forest Resources Management, University of Ibadan.)

BLACK, C. C. (1971) Ecological implications of dividing Plants into groups with distinctive photosynthetic production capacities. *Advances in Ecological Research*, Vol. 7, pp. 87–114.

CHILD, G. S. (1974) *An ecological survey of the Borgu Game Reserve* (F.A.O., Rome.)

EGGELING, W. J. (1964) A nature reserve management plan for the Island of Rhum, Inner Hebrides. *Journal of Applied Ecology*, Vol. 1, pp. 405–19.

GEERLING, C. (1975) *The 1:50 000 vegetation map of Borgu Game Reserve.* (F.A.O., Rome.)

HOWELL, J. H. (1968) The Borgu Game Reserve of northern Nigeria Part 1. *Nigerian Field*, Vol. 33, pp. 99–116.

ITA, E. O. (1972) *A report of the fisheries survey of the Olli River in Borgu Game Reserve.* (F.A.O., Rome.)

JANZEN, D. H. (1975) *Ecology of plants in the tropics.* (Edward Arnold, London.)

LAMOTTE, M. (1975) The structure and function of a tropical savannah ecosystem. In *Tropical Ecological Systems.* (Golley, F. B. (Ed.) and Medina E.). (Springer–Verlag, New York), pp. 179–211.

WALTER, M. W. (1967) Length of the rainy season in Nigeria. *Nigerian Geographical Journal*, Vol. 10, pp. 123–8.

25

SOME PROBLEMS ENCOUNTERED IN THE FIELD STUDY OF THE GRASSCUTTER (*Thryonomys swinderianus*) POPULATION IN GHANA

Emmanuel O. A. Asibey

Department of Game and Wildlife,
Accra, Ghana

INTRODUCTION

The grasscutter *(Thryonomys swinderianus)* is of great economic importance in Ghana both as an agricultural pest and as a source of animal protein, very well acceptable to all classes of people. The nutritive value of this animal is high and it compares very favourably with beef, mutton and pork which is available on the Ghanaian market (Asibey, 1974a). The rabbit *(Oryctolagus cuniculus)* is reported to have the highest (20.7 per cent) protein yield of all meat eaten in Britain (Ministry of Agriculture, Fisheries and Food, 1973). My preliminary analysis of grasscutter meat has indicated that the protein content (22.7 per cent) of the grasscutter in Ghana is higher than that of the rabbit.

Attempts were made to study the animal both in captivity and in the wild. The captive stock proved successful and indicated that it is an animal which can easily be domesticated like its distant relative the guinea pig *(Cavia porcellus)*. Consequently, its domestication on large scale is being investigated in Ghana.

The primary purpose of the domestication is to help interested people to breed grasscutter for their own consumption or for cash income or both. Consequently, right from the onset, the public's attention was drawn into the project. They were taught to use the same rearing techniques and feed as used for those bred in the laboratory. They were visited at least once a week besides special visits on request. Breeding boxes and animals were given to interested people who took care of their animals. It was found that the domestication was technically feasible and that the current market price makes it a worthwhile venture in both urban and rural areas.

Den Hartog and de Vos (1973) indicated that the management of wild populations of grasscutter is more likely to find a wider application in West Africa than the domestication of the animals and therefore more attention should be given to the management of wild population than to domestication.

214

Ajayi (1974) agreed with den Hartog and de Vos. Unless wild grasscutter population can be studied, habitat manipulation to enhance grasscutter population build up for the purpose of meat production is not likely to be the answer to the need for commercial production of grasscutter meat on a sustained yield basis.

METHODS

Sites were chosen in the Shai Hills Game Production Reserve and the Mole National Park for the study of unexploited population, and in Apam area for the study of exploited population, to find out the relationship and effects of habitat manipulation and continuous exploitation on the numbers and population turnover, over a period of three years, using capture-recapture techniques.

Three techniques were used for catching the grasscutters namely (a) cage trapping, (b) combined pen with cage trapping, (c) netting with drivers. The catch were to be subsequently marked and released for population studies. Baits used were sugar cane (*Saccharum* sp.), cassava *(manihot utilissima)*, maize *(Zea mays)*, rock salt and sandstone soaked in human urine were used as baits in the traps but they were not taken by the animals. When none of the three techniques was effective, a fourth technique, inspectional search with lights, was tried but this also was not successful. A brief description of the techniques follows below:

(a) Cage trapping

Three kinds of traps were used for this study. The box trap used by Dunnet in the study of quokka *(Setonix brachyurus)* (Dunnet, 1956) was modified for this study, to allow previewing of the catch before handling. The modified box traps were made of aluminium (because it is cheap and light) with poison free paint. They were large enough to avoid damage to the animal as the door closed behind it and provided protection against weather and predators. The trap was used in various ways but it was unsuccessful.

The second type of trap was the Havasack trap. This was made of chicken-wire. It is rounded at the mouth and flattened at the end where it is sealed off. A funnel shaped piece is inserted into the mouth end to allow the animal to enter but prevent it from getting out.

This trap caught one grasscutter but there was no other catch thereafter. The animal did not wound itself or attempt to gnaw its way out.

(b) Pen with cage trapping

This technique was tried at Shai Hills Game Production Reserve and in Apam area. At Shai, three plots of varied sizes were fenced with chicken-wire. One of the plots was cleaned and planted with maize, but due to a bad season the maize did not grow well. At Apam, farms on isolated hills were fenced. Entrances were made in the fence to allow animals into the enclosures and crab-traps were found near and around each of the pens, no animal entered any of the traps.

Feeding/drinking points were established at Shai using maize (fresh, dry, salted) cassava, sugar cane, common salt and human urine as bait. Water was provided in plates and tins. The points were inspected regularly for signs of grasscutter activity in the locality.

The idea was that when the animals were known to use the facilities provided, they would be netted, marked by ear tags and toe clipping and then released. A hide-out was to be built from where the site(s) could be observed and the activities of the animals studied. The points established were never used by the animals and therefore the method was abandoned.

(c) Netting and drivers

Another method which was tried involved driving the animals into nets. This was tried in the Mole National Park. Three plots were chosen, one in a swampy area, another on a slope of a ridge down into the edge of a swamp, and the third on a dry hill-top far removed from water. Grasscutter feeding sites were observed in all the plots.

Within the selected areas a 100 yards nylon net of mesh 10×10 cm was set along a path or a marked track. Beaters (ranging from 10–20 in numbers) were lined up at distances of 100 m from the net. They beat the vegetation and made noises to drive the animals towards the net. Other people stood behind the net to watch animals getting caught in it.

In all the plots grasscutters were seen in the operation but none were caught in a net. They moved ahead of the beaters but at the net, they would halt, freeze, and move slowly along the net line for about 1 m and back out in the same, slow manner to meet the beaters who did not notice them.

Large mammals: bushbuck *Tragelaphus scriptus* Pallas and kob *Kobus (Adenota) kob* Erxleben encountered ran ahead of the beaters and broke through the net.

DISCUSSION

The grasscutter has been mentioned in most natural history books on the mammals of Africa south of the Sahara and although it is recognized everywhere as both a pest and a source of meat to people (Asibey, 1974b) I have not come across any work on the population study of the animal. Nor does its biology seem to have been studied (Ewer, 1960).

The animal, which originates from the savannah country (Booth, 1959; Asibey, 1974b), has successfully penetrated the forest zone. The occurrence of this animal whose meat is acceptable to most people in protein-deficient Africa (Ajayi, 1971, 1974; Asibey, 1974b; Cockrill, 1967; Haisel, 1971; Olayide, 1973; Ovington, 1963) makes it an important animal to be thoroughly studied.

The exploitation of wild populations has two aspects: as a measure of pest control and as a source of animal protein. In order to utilize wild populations economically and avoid extermination it is necessary to be able to assess the size of the population from time-to-time to determine the crop to be taken and the effectiveness of both management and cropping techniques.

It is therefore most essential that suitable techniques be developed for the study of wild populations. At the moment, such knowledge is lacking and it is essential that greater attention is paid to the development of such field techniques.

Ewer (1969) observed that the grasscutter is conservative in its feeding habits. This may account for the ineffectiveness of baits for trapping. This raises a serious problem of finding suitable baits in the study of wild grasscutter population. The animal is naturally commonest in habitats where its food and cover are abundant. The use of such food items in the same habitat as baits will be of no use as there will be nothing special about the bait to attract them. Although theoretically it may be assumed that the management of wild populations is the best approach to the exploitation of this species as a source of animal protein techniques are yet to be developed whilst demand for this source of meat grows rapidly.

The available knowledge shows that there is no apparent reason why the domestication of the grasscutter should not continue, whether or not wild populations can be managed through 'simple habitat manipulation' (Den Hortog and de Vos, 1973). The two ways of approaching the problem for meat production seem to be complementary not substitute.

ACKNOWLEDGEMENT

I would like to thank the Government of the Republic of Ghana for sponsoring my participation in this symposium.

REFERENCES

AJAYI, S. S. (1971) Wildlife as a source of protein in Nigeria. *Nigerian Field*, Vol. 36, pp. 115–27.

AJAYI, S. S. (1974) Giant Rats for meat and some taboos. *Oryx*, Vol. 12, pp. 379–80.

ASIBEY, E. C. A. (1974a) The Grasscutter, *Thryonomys swinderianus* (Tem.) in Ghana. *Symp. Zool. Soc. Lond.*, Vol. 34, pp. 161–70.

ASIBEY, E. C. A. (1974b) Some ecological and economic aspects of grasscutter, *Thryonomys swinderianus* Temminck *(Mammalia, Rodentia, Hystricomorpha)* in Ghana. (Unpublished Ph.D. thesis, University of Aberdeen.)

BOOTH, A. H. (1959) On the mammalian fauna of the Accra Plains. *J. W. Afr. Sci. Ass.*, Vol. 5, pp. 26–36.

COCKRILL, R. W. (1967). (ed.): *World Protein Hunger — The Role for Animals*. (FAO).

DEN HARTOG, A. P. and DE VOS, A. (1973) The use of rodents as food in Tropical Africa. *FAO. Nutr. Newsl.*, Vol. 11, pp. 1–14.

DUNNET, G. M. (1956) A population study of the Quokka *Setonix brachyurus* Quoy and Gaimard (Marsupiala). *Journal of Mammalogy*, Vol. 44, No. 1.

EWER, R. F. (1969) Form and function in the Grasscutter *Thryonomys swinderianus* Tem. (Rodentia Thryomidae). *Ghana Journal of Science*, Vol. 9, pp. 131–49.

HAISEL, K. A. (1971) Ecological problems of Agricultural Production in West Africa: In Ofori, I.M. (1973) (Ed.). *Proceedings of Conference on Factors of Agricultural Growth in West Africa*. (I.S. S.E.R. Legon).

MINISTRY OF AGRICULTURE, FISHERIES AND FOOD (1973) *Commercial Rabbit Production Bulletin 50*. (HMSO, London.)

OLAYIDE, O. S. (1973) Some Aspects of Beef Production in Nigeria: In Ofori, I.M. (1973) (Ed.). *Proceedings of Conference on Factors of Agricultural Growth in West Africa 1971*. (I.S. S.E.R. Legon.)

OVINGTON, J. D. (1963) (Ed.). *The Better Use of the World's Fauna for Food*. (The Institute of Biology, London.)

26

THE ROLE OF THE DEPARTMENT OF FOREST RESOURCES MANAGEMENT, UNIVERSITY OF IBADAN, IN TEACHING AND RESEARCH IN WILDLIFE MANAGEMENT

L. Roche

Department of Forest Resources Management,
University of Ibadan†

I imagine that there will be no disagreement about the need for full university type courses for professional staff of Wildlife Departments. There is room for discussion, however, as regards the length and structure of such courses, of the extent to which they should be specialized, and as to where they should be offered. (Edney, 1963)

INTRODUCTION

It is not universally accepted that foresters are people who in certain circumstances may play a major and beneficial role in the general field of wildlife conservation and management. For example, at the Seventh Biennial Conference of the West African Science Association held in the University of Ibadan in 1970 the following statement, with little qualification or supporting evidence was made: 'Nigeria is now divided into twelve states, each with its own Forestry Department responsible for the wildlife of the state. This is clearly an unsatisfactory condition, and one that is likely to be detrimental to the future of wild animals in the Federation' (Happold, 1971). The author goes on to advocate a centralized administration staffed by zoologists and ecologists. There is no mention of the possible role of government departments of forestry in wildlife conservation and management in Nigeria despite the fact that virtually all major achievements in conservation and management in this field to date have stemmed from the activities of federal and state forestry departments.

The implication of this point of view is that the University's Department of Forestry will also be 'detrimental to the future of wild animals in the

†Department of Forestry and Wood Sciences, University College of North Wales, Bangor, United Kingdom.

Federation', for, after all, this department is the only source of locally educated foresters in the Federation at the moment, and its graduates staff the government's forestry departments in all twelve states as well as the two Federal Departments of Forestry. Of course, the Department also caters to the needs of a significant number of English-speaking countries throughout Africa for forestry graduates.

It is not the purpose of this brief paper to state with equal firmness, the opposite view to that quoted above, but firstly to present an outline of the growth of the Department's interest in this field; secondly to examine the capability of the department in relation to teaching and research in wildlife conservation and management; and thirdly to stimulate discussion at this gathering of experts on the possible role of a major West African university and its diverse departments in teaching, research and development in the very broad field of wildlife conservation and management.

I say university rather than department, for I do not wish to imply that it is desirable or feasible for the Department of Forest Resources Management of the University of Ibadan to have a monopoly interest in wildlife management problems, though I take it as indisputable that it has a major role to play in this field, and that it is now in a position to do so.

The traditional view of the forester as a producer and processor of cellulose is increasingly irrelevant and, since the education of foresters in modern African states has been the subject matter of a separate paper (Roche, 1975), it is not proposed to discuss it further here. The paper referred to, however, particularly in relation to the Department of Forest Resources Management at Ibadan, is rather essential background material to what is presented here.

HISTORY

At its inception in 1963 the department was named 'Department of Forestry' though even then its mandate included range and wildlife management, and it was anticipated that in time the allied resource management disciplines, watershed management and the management of fresh-water fisheries would be added. By 1974 the Department had considerably increased its teaching and research responsibilities. This increased responsibility, together with the growth and expression of a coherent philosophy of integrated natural resource management within the Department, persuaded the University's Senate to accede to a departmental request for a change of name, and in due course the Department was called 'Department of Forest Resources Management'.

This change of name, therefore, reflected two important developments within the Department; the first being the Department's clear and articulated commitment to the principles of integrated natural resources management, and second the considerable increase in the number and diversity of academic backgrounds of the senior staff of the Department, which allowed it to give concrete expression to this commitment.

CAPABILITY

Before presenting factual information about the capability of the Department in teaching and research in wildlife management, it is perhaps worth examining some views on what constitutes expertise appropriate to the wise management of a wildlife resource.

Even a cursory reading of the proceedings of the Arusha conference held in 1961 and of the EAAFRO conference held in Nairobi in 1967 indicates three principles of fundamental importance with regard to the wise management of a wildlife resource. The first is that unity and cohesion are given to teaching, research and development in wildlife management when its ecological basis is clearly recognized and understood. Secondly, a preoccupation with zoological problems to the near exclusion of habitat problems leads to disaster. Thirdly, there must be the close liaison between researchers and teachers on one hand and actual managers of land and animals on the other. A research élitism which has scant respect for, or knowledge of, the day-to-day difficulties of men in the field may be intellectually satisfying but it accomplishes little in modern African States.

In the proceedings of the Nairobi symposium the following statement is made in relation to the lack of information on the ecology of Miombo country in East Africa where game reserves were being set up:

> We have seen the technical and scientific experts operate in little boxes, and the Game Division seems to operate in the long grass, and it seems never the pair will meet. It is rather disappointing to open scientific journals and see that they read rather like the Kinsey Report on the larger mammals. The type of information I think we basically need is population dynamics, predator/prey relationships and possibly very detailed ecological papers on these subjects related to the 'miombo' type of vegetation.
> The sex life of the impala may be very interesting, but as long as we don't know what they eat they may have no future sex life at all.   (Rees, 1968)

I am not in a position to assess the validity of Glover's conclusions at the Arusha conference (Glover, 1963) that 'Unfortunately East Africa has lost the best part of 20 years because the importance of coordinated ecological work was not recognized early enough', except to say that there is nothing in the published proceedings of the conference which indicates that he was wrong.

Basic zoological and related problems are in my mind best dealt with, at a university level, in departments of zoology and veterinary science, and similarly basic botanical problems are more appropriately dealt with in departments of botany. The accumulated results of research in these departments, or at least the immediately relevant portion of it, must be incorporated in plans for the management of renewable natural resources. A department of forest resources management on the other hand, must be concerned with problems of management and management systems, and, therefore, can play a central role in ensuring two things: first that relevant problems in wildlife management are brought to the attention of these other departments, and secondly, that relevant research results from these departments are synthesized and codified in management plans, including the management of wildlife.

Does such a department, then, have no research problems of its own? Not at all, for if it is to push forward the frontiers of knowledge of management systems then it will be continually concerned with management problems. Furthermore, if basic research data is missing and not forthcoming from the departments referred to above it may be forced to seek it itself.

In any event, if it is a vigorous department there will always be a tendency to push back into basic research problems. This is not a bad thing, for it is this tendency which ensures that a department principally concerned with resource management links up with the other departments referred to above.

There are, however, relatively clearcut problems in management which can be appropriately tackled by a department preoccupied with teaching and research in natural resources management, for example, in regard to wildlife, such as those lucidly set out by Lamprey (1963):

> The information which is required for a proper understanding of the factors involved in a policy of wise conservation and management can be described in very broad terms. We need to know how many animals are present in the area with which we are concerned and what are their reproductive and survival rates. What species are present and in what numbers? How do they make use of the country? Do they remain in one area throughout the year or do they move over considerable distances? How are they affected by the seasonal variations of the climate and the changing availability of water supplies? What types of vegetation does each species select for its habitat and what is the significance to them of the different plants and plant associations? Is the habitat stable or is it changing? These and many other questions demand answers.

The manner in which a university department of forest resources Management could seek answers to nearly all of these questions for a particular game reserve is given in Bulletin 3 of this Department entitled *The Vegetation of the Yankari Game Reserve* (Geerling, 1973).

Ideally, therefore, a university department concerned with teaching and research in renewable natural resources management should co-operate closely with, and occupy that ground between, university departments of zoology, botany, veterinary science, animal husbandry, and on those government departments charged with the actual management of renewable natural resources. This, in fact, is how the Department of Forest Resources Management at Ibadan is presently functioning.

The Department has now a senior staff of 20 including ecologists, zoologists, foresters, and an engineer as well as an array of expertise represented by visiting staff. It has complete access for research and teaching purposes to all the forest and game reserves of the Federation including the Forest Reserve at

Ijaiye which is 193 sq. km and only 38 km from Ibadan, and is used, thanks to the Government and Chief Conservation of Forests of the Western State, as the Department's field station. The array of disciplines in the Department ensures a degree of interdisciplinary research and teaching impossible in the more traditional university departments of forestry, and its growing competence in resources biometry, and the use of aerial photographs and remote sensing generally, places it in a position of strength in relation to habitat assessment (see appendix 1 for a list of research projects in wildlife management presently underway in the department).

THE FUTURE

As will be seen from the appendix, the Department is clearly in a strong position academically, and can, with confidence, accept the challenge of teaching and research in wildlife management at university level that is now being imposed by the demands of Nigeria's 14 Departments of Forestry, and by English-speaking Africa generally. It is already offering postgraduate training in this field. All its undergraduates are offered courses in wildlife management, and consequently are fully aware of the economic and aesthetic importance of this resource when they enter State and Federal departments of forestry as foresters. Those with postgraduate training are usually offered positions appropriate to their qualifications in the game sections of these departments.

   A number of questions remain as to future developments in the Department in this field. It is outside the scope of this paper to examine these questions in detail, but in order to solicit a feedback to the Department from this symposium it may be well worthwhile to pose them:
(a) Should the Department now offer a diploma in wildlife management? The department already offers a diploma — lasting one year — in forestry usually taken by graduates in Zoology, Botany, or the allied biological disciplines, who wish to obtain a professional qualification. If so what should its content be?
(b) Should the department offer a first degree in wildlife management — at present its only first degree is a B.Sc. in forestry — or should it simply offer a specialization in wildlife management to those final year forestry students who wish to take it?
(c) Should, in the long run, the wildlife section within the Department form a fully fledged department of its own within a faculty of natural resources management as is the situation in a number of other countries outside Africa where wildlife is an important resource?

   As already indicated, it is outside the scope of this paper to attempt answers to these questions. But to my colleagues concerned with wildlife management at the University of Ibadan I wish to say that at the present time on this continent with its vast human population, its vast ecological potential, and the possibilities for management in perpetuity of its natural ecosystems, I say you cannot be too ambitious. Within this context a fully-fledged University

Department of Wildlife Management at Ibadan is a practical and desirable objective.

The day is passed when young Africans interested in wildlife management were compelled to seek appropriate training overseas in temperate countries. They will, and rightly so, continue to travel abroad for this purpose, but a choice is now available to them, if they so wish, to obtain their training in a more appropriate setting, that is, within tropical ecosystems, and within the cultural environment of an African Nation. I am convinced, that at least for the English-speaking countries of West and Central Africa the Department of Forest Resources Management, University of Ibadan, is now ready to offer that choice.

REFERENCES

EDNEY, E. B. (1963) Education and training of staff. In IUCN Publication New Series No. 1. (IUCN Morges).

GEERLING, C. (1973) The vegetation of the Yankari Game Reserve. Bulletin 3 (Department of Forestry, University of Ibadan.)

GLOVER, P. E. (1963) Factors of the habitat. In IUCN Publication New Series No. 1. (IUCN Morges).

HAPPOLD, D. C. D. (1971) A history of wildlife conservation in Nigeria, and thoughts for the future. Wildlife Conservation in West Africa. (IUCN Publication No. 22. Morges).

LAMPREY, H. (1963) The survey and assessment of wild animals and their habitat in Tanganyika. In IUCN Publication New Series. No. 1. (IUCN Morges).

REES, A. F. (1968) Proceedings of the symposium on wildlife management and land use. *African Agricultural and Forestry Journal*, (Special issue), p. 163.

ROCHE, L. (1975) Major trends and issues in forestry education in Africa: a view from Ibadan. *Commonwealth Forestry Review*, Vol. 54, pp. 166–75.

APPENDIX 1

I. Domestication of African wildlife as a source of meat.
   1. The biology and domestication of the African giant rat (*Cricetomys gambianus* Waterhouse).
   2. The biology and domestication of the grasscutter (*Thryonomys swinderianus* Temminck).
   3. Management of the African giant snail *Archachatina marginata* as a source of meat in Southern Nigeria.
   4. The biology and domestication of the bushfowl *(Francolinus bicalcaratus)*.
II. Habitat and Population Studies.
   1. Habitat utilization and population dynamics of the free-ranging Fulani cattle in Northern Nigeria.

      The objectives of the project are to investigate the population structures, and reproductive performance of the Fulani cattle, their various grazing patterns under nomadic, sedentary (re-settlement) and modern ranching conditions in order to formulate a management policy that would ensure an optimum form of land use.
   2. Effects of burning and grazing on the structure and productivity of savannah grassland in Borgu Game Reserve.

      This project examines the effects of fire and grazing on the herbage productivity and mineral cycling in Kob *(Kobus kob)* grazing areas of the reserve, analysis of structure and productivity of the major grass species in the three principal vegetation types, and changes in floristic composition and habitat trends in response to different fire treatments.
   3. A comparative study of the social behaviour and ecology of Western hartebeest *(Alcelaphus buselaphus major)* and Roan antelope *(Hippotragus equinus)* in Borgu Game Reserve.

      The objective of the study is to assess the distribution, abundance and herd structure of Roan antelope, and Western hartebeest in Kainji Lake National Park. The information is required for the formulation of a management policy for important wildlife species of Kainji Lake National Park.
III. Management Plans.
   1. Formulation of Management Policy for Kainji Lake National Park.

225

The objective is to synthesis the existing information on the fauna and flora, the soils, topography, and drainage system of Borgu Game Reserve, the present management problems, in order to formulate a policy for the future effective management of the reserve.

LIST OF PARTICIPANTS

Abolude, F., Federal Department of Forestry, Wildlife/Conservation Division, P.M.B. 5011, Ibadan

Adekunle, A. O., M.A.N.R., Forestry Division, P.M.B. 5007, Ibadan

Adetona, J. G., Deans Office, Faculty of Agriculture, University of Ibadan

Adewetan, T. A., Federal Department of Forest Research, Wildlife Section, Ibadan

Afolayan, T. A., Department of Forest Resources Management, University of Ibadan

Aire, T. A., Veterinary Anatomy and Physiology, University of Ibadan

Ajayi, S. S., Department of Forest Resources Management, University of Ibadan

Akande, M. (Mrs), Department of Forest Resources Management, University of Ibadan

Akinsanmi, K., Department of Forest Resources Management, University of Ibadan

Aladejana, K., Federal Department of Forestry, P.M.B. 5011, Ibadan

Amika, E. B. O., Department of Forestry, Calabar

Anadu, P. A., Department of Zoology, University of Ibadan

Ayeni, J. S. O., Kainji Lake Research Project, New Bussa

Amon, B. O. E., A.R.C.N.. Ibadan

Biala, M., Nnamdi Azikwe Hall, University of Ibadan

Brinckman, W. L., Ahmadu Bello University, Zaria

Deluew, P. N., Department Animal Science, Ahmadu Bello University, Zaria

Danisa, R., Ministry of Agriculture and Natural Resources, Forestry Division, Benin City, Bendel State

Dipeolu, O. O., Department of Veterinary Pathology, University of Ibadan

Egunjobi, J. K., Department of Agricultural Biology, University of Ibadan

Faniyi, O., M.A.N.R., Forestry Division, P.M.B., 5007, Ibadan

Funmilayo, O., Department of Agricultural Biology, University of Ibadan

Gadzama, M., Department of Biological Sciences, Ahmadu Bello University, Zaria

Hall, P., Ministry of Natural Resources, Forestry Division, Maiduguri

Idrisu, A., Forestry Division, Ministry of Animal Health and Forest Resources, Sokoto

Inogbo, O., Department of Forest Resources Management, University of Ibadan

Isichei, O., Department of Biological Sciences, University of Ife, Ile-Ife

Isoun, T. T., Veterinary Pathology Department, University of Ibadan

Jaiyesimi, A. K., Forestry Division, M.A.N.R., Ibadan

Jia, J., Game Preservation Unit, Forest Division, Bauchi

Joseph, O. T., Museum of Natural History, University of Ife, Ile-Ife

Joshi, P. N., Ministry of Coops, Forestry and Rural Development, P.M.B. 3097, Kano

Kapu, M., Biological Sciences, Ahmadu Bello University, Zaria

Kio, P. R. O., Department of Forest Resources Management, University of Ibadan

Lapai, A. M., Ministry of Animal Health Forest Resources, North Western State, Sokoto

Lasan, A. H., Game Preservation Unit, Bauchi, Nigeria, Forestry Division, M.N.R., Maiduguri

Lowe, Federal Department of Forest Research, P.M.B. 5054, Ibadan

227

Milligan, K. R. N., Department of Forest Resources Management, University of Ibadan
Mordi, R., Ministry of Agriculture, Benin City
Mustafa, M. H., P.M.B. 3097, Kano
Ogunsanmi, A. O., Federal Department of Forest Research, Ibadan
Ojo, M. O., Department of Veterinary Pathology, University of Ibadan
Okali, D. U. U., Acting Head, Department of Forest Resources Management, University of Ibadan
Okoye, I. G., Department of Forest Resources Management, University of Ibadan
Okigbo, B. N., Chairman, Agricultural Research Council of Nigeria, Moor Plantation, Ibadan
Olayide, S. O., (Representing the Vice-Chancellor, University of Ibadan) Dean, Faculty of Agriculture and Forestry, University of Ibadan
Oladejo, J. O., M.A.N.R., Forestry Division, P.M.B. 5185, Ibadan
Oladipo, F. F., Oladipo, M.A.N.R., Forestry Division, P.M.B. 5007, Ibadan
Olowo-Okorun, M. O., Department of Veterinary Anatomy and Physiology, University of Ibadan
Olufajo, O. O., Department of Forest Resources Management, University of Ibadan
Onadeko, S. A., Department of Forest Resources Management, University of Ibadan
Oyatogun, M. O. O., Department of Forest Resources Management, University of Ibadan
Oni, I. O., Department of Forest Resources Management, University of Ibadan
Oregbeme, T. O., M.A.N.R., Udo Gand Complex Ubiaja, Bendel State
Oseni, A. M., Director, Federal Department of Forestry, P.M.B. 5011, Ibadan
Oyenuga, V. A., Deputy Vice-Chancellor and Head of Department of Animal Science, University of Ibadan
Oyewole, E. I., Department of Forest Resources Management, University of Ibadan
Saba, A. R. K., Federal Department of Forestry, Ibadan
Sanford, W. W., University of Ife, Ile-Ife
Sanwo, S. K., Department of Forest Resources Management, University of Ibadan
Sarkar, N., Ministry of Coops, Forestry and Rural Development, P.M.B. 3097, Kano
Steinbach, J., University of Ibadan
Taussig, Faculty of Veterinary Medicine, Ahmadu Bello University, Zaria

FOREIGN PARTICIPANTS

Allo, A. A., Garoua Wildlife College, B.P. 271, Garoua, Cameroun Republic
Andrianiriana, G., Direction des Eaux et Forêts, Tananarive, Madagascar
Asibey, E. O. A., Department of Game and Wildlife, Box M239, Accra, Ghana
Barber, K. B., B.P. 1080, Bangui, Central African Empire
Bete, T., Controlean Eaux et Forêts Natitingou, Cameroun Republic
Bosch, M. L., B.P. 271, Ecole de Faune, Garoua, Cameroun Republic
Carver, D. A., Carver, B.P. 971 Cotonou, Benin Republic
Chabwela, H. N., Wildlife and Parks B.P.1 Chicanga, Zambia
Child, G. S., Via Delle Termi di Caracalla, Rome 00100, Italy
Christenson, B., B.P. 537, Ouagadougou, Upper Volta
Cobb, S. M., Department of Zoology, Oxford University, UK
Edroma, E. L., Uganda Institute of Ecology, P.O. Box 22, Lake Katwe, Uganda
Forcier, J. P., B.P. 537, Ouagadougou, Upper Volta
Geerling, C., B.P 271, Caroua, Cameroun Republic
Groaua, N'Dein, Secretariat I'rata our Paces Nationaux (SEPN), Ivory Coast
Gwynne, M. D., NNDP/FAO Habitat Utilization Project, P.O. Box 30218, Nairobi, Kenya

Hechtel, J., P.M.B. 5, Bamenda N.W.P., Cameroun Republic
Hudson, M. R., B.P. 537, Ouagadougou, Upper Volta
Jeffery, S., c/o P.O. Box 2075, Monrovia, Liberia
Kaghan, B., B.P. 537, Niamey, Republic of Niger
Kavanagh, M., c/o E L'Ecole de Faune, Garoua, Cameroun Republic
Koivisto, I., Helsinki Zoo, 00510 Helsinki 57, Finland
Kongoro, Z. O., Ministry of Tourism and Wildlife, Box 30027, Nairobi, Kenya
Lafforest, J. De., c/o 29 Villiers Street, London WC2, UK
Larson, T. J., B.P. 971, Cotonou, Benin Republic
Latakpi, Co-Directeur Project CAF/72/010, Studies Frieluim. Bangui B.P. 830, Centraficaine, Central African Empire
Lavieren, van L. P., Ecole de Faune B.P. 271, Garoua, Cameroun Republic
Levy, R., Directeur des Chasses B.P. 830 Bangui, Central African Empire
Luketa, S., c/o Direction des Faune et Forêts B.P. 8722 Kinshasa/Gombe, Zaire
Mahamat, A., Homologurs Project Assistance, Parcs Nationaux Garoua, Cameroun Republic
Mankoto, M.M., Attaché de Recherches a IZCN, B.P. 4019 Kinshasa, Zaire
Mapunda, W. J., Tanzania Game Division, P.O. Box 1994, Dar-es-Salaam, Tanzania
Mbuvi, D. M., Kenya Wildlife Project, Box 30559, Nairobi, Kenya
Minner, E. D., B.P. 2 Bouna, Ivory Coast
Modha, M. L., Kenya Game Department, Box 40241, Nairobi, Kenya
Monfort, D., P.O. Box 965, Nigali, Rwanda
Montague, W. C., S/C Corps de la Paix, B.P. 537, Niamey, Niger
Myers, N., FAO Box 1625, Accra, Ghana
Ndiawar, D., Eaux et Forêts, B.P. 1813, Dakar, Senegal
Ngog, N. J., B.P. 271 Garoua, Cameroun Republic
Ntiamoa-Baidu, Y. (Mrs) Department of Game and Wildlife Box M239, Accra, Ghana
Nmiamed, E. K., Forestry and Wildlife Project, SOM/72/012, Somali Dam, Somali
Oguama, Direction des Eaux et Forêts et Chasses, B.P. 397 Cotonou, Benin Republic
Pierrent, B. P. 1942, Kisangani, Zaire
Riney, T., 31 Queens Crescent, Edinburgh EH9, UK
Roche, L., Department of Forestry and Wood Science, University College of North Wales, Bangor, North Wales, UK
Rodgers, W. A., Tanzania Game Division, Box 1994, Dar-es-Salaam, Tanzania
Sale, J. B., Department of Zoology, University of Nairobi, P.O. Box 30197, Nairobi, Kenya
Sanogho, N., Directeur des Parcs Nationaux de la République du Mali, Mali
Spinage, C. A., UNDP, B.P. 872, Bangui, Central African Empire
Stark, M.A., Project des Parcs Nationaux, B.P. 237, Garoua, Cameroun Republic
Sutterfield, R. W., B.P. 623, Bouafie, Ivory Coast
Tentchou, J., Ecole de Faune, B.P. 271, Garoua, Cameroun Republic
Vanpraet, C., B.P. 237, Garoua, Cameroun Republic
Wagner, M. D., B.P. 11, Tanguieta, Benin Republic
Wise, W. A., P.O. Box 1062, Tahee City, California 95730, USA
Woodford, M. H., P.O. Box 30559, Nairobi, Kenya

INDEX

Acacia woodland, 207–8
Acinonyx jubatas, 195
Adenota kob, 167, 168
Aepyceros melampus, 34, 46, 47, 73, 74, 75, 76, 77, 78, 79, 156
Aerial censuses, 50–1
Aerial photography, 2–4
Aerial surveys, 1–7, 8–18, 64, 70
African buffalo, *see* Buffalo
African giant snail, 225
African swine fever, 59, 63
Afzelia woodland, 207, 208, 210
Age prediction in ungulates, 33–40
Age structure, 27–30, 64
Alcelaphus buselaphus, 53, 54, 73, 76, 77, 86, 90, 91, 92, 123, 125, 156, 195, 225; as lions' prey, 43, 44, 45, 47
Antelope, *see* Roan antelope
Anthrax, 58, 59
Anubis baboon, *see* Baboon
Archachatina marginate, 225
Aristida adoensis, 178, 179, 180
Arthropod vectors, 56, 57, 58, 59, 60, 62, 63, 64
Arvicanthis niloticus, 129, 131, 137, 138, 139, 140

Baboon, 47, 63, 86, 90, 91, 195
Baoulé National Park, Mali, 194–204
Bilharzia, 63
Black/grey rat, 129, 130–1, 133, 137, 138, 139, 140
Blood of giant rat, 142, 143–9
Blue duiker, *see* Duiker, common
Bothriochloa insculpta, 178, 179, 180
Bouba Ndjida National Park, Cameroun Republic, 51–5
Brachiaria decumbens, 180
Browse, 101–9, 110–21; in Baoulé National Park, Mali, 199, 201; mineral content of, 102–3; protein con-

tent of, 102, 104–6, 107; species of, 112, 115, 116, 117, 118, 119
Brucellosis, 59
Bubal hartebeest, *see* Hartebeest
Buffalo, 34, 37, 38, 64, 73, 74, 77, 78, 80, 81, 86, 90, 91, 92, 158, 195, 196; as lions' prey, 43, 44, 45, 46, 47, 48
Burkea africana/Terminalia avicennioides woodland, 207–8, 210
Burning, 96–7, 151–9, 160–5, 166–75, 189, 198, 208, 210, 212, 225; erosion and, 161–2; grassland and, 153, 154, 155, 156, 157–8, 166–75; land surface and, 152; litter and, 169–70, 173; pests and, 174; plant productivity and, 176–85; soil and, 151, 161–2, 170; standing crop and, 170, 171, 172, 181; uses of, 161; vegetation and, 152, 153, 154–8, 161, 162–4; water and, 151; wildlife and, 152, 153, 163, 164
Bushbuck, 47, 195
Bushfowl, 225
Bush meat, 1, 127, 133, 134, 214; African giant snail as, 227; bushfowl as, 225; giant rat as, 225; grass cutter as, 214–15, 225
Bushpig, 47

Calcium, 102, 103
Cane rat, 28–9, 129, 132, 133–4, 138, 139, 214–17, 225; domestication of, 214–15; trapping of, 215–16
Catarrhal fever, 60
Cattle, 225; browse preferences of, 101–9, 110–21; diurnal activity of, 111, 112, 113; zebu, 198
Cephalophus rafilatus, 195
Cercopithecus aethiops, 195
Cheetah, 195
Chloris gayana, 179, 180; *C: pycnothrix,* 178, 179

231

For Product Safety Concerns and Information please contact our EU
representative GPSR@taylorandfrancis.com
Taylor & Francis Verlag GmbH, Kaufingerstraße 24, 80331 München, Germany